Engineering Drawing and Product Manufacturing Information with 3D Models

Instructing readers on both basic and complex drawing techniques, *Engineering Drawing and Product Manufacturing Information with 3D Models* is an instructive reference to the use of 3D computer models in modern industry.

This book provides a comprehensive guide to the use of 3D computer-aided design (CAD) models for communicating design intent, mainly through the adoption of International Standards Organization (ISO) methods for describing shapes and for depicting dimensions and tolerances on drawing sheets or other product manufacturing information–based (PMI) media. It describes the fundamentals of computer numerical control (CNC) and the generation of 3D printing and additive manufacturing models as well as basic fabrication specifications. Common file types used to store, share and transfer media are described in some depth.

Engineering Drawing and Product Manufacturing Information with 3D Models will be of interest to students and engineers working with 3D models in fields including, but not limited to, mechanical, electrical, industrial and biomedical engineering, along with materials and computer science.

Engineering Drawing and Product Manufacturing Information with 3D Models

ISO Geometrical Product Specification Standards

Frank Mill
CEng BSc(hons) PhD FIMechE PFHEA

CRC Press
Taylor & Francis Group
Boca Raton London New York

CRC Press is an imprint of the
Taylor & Francis Group, an **informa** business

Designed cover image: Frank Mill

First edition published 2025
by CRC Press
2385 NW Executive Center Drive, Suite 320, Boca Raton FL 33431

and by CRC Press
4 Park Square, Milton Park, Abingdon, Oxon, OX14 4RN

CRC Press is an imprint of Taylor & Francis Group, LLC

ISBN: 978-1-032-49698-6 (hbk)
ISBN: 978-1-032-49700-6 (pbk)
ISBN: 978-1-003-39507-2 (ebk)

DOI: 10.1201/9781003395072

Typeset in Times
by Apex CoVantage, LLC

Contents

Preface xv
Acknowledgements xvii
Author xix
Introduction: 3D Modelling and Downstream Information xxi

Chapter 1 Basic Drawings: Common Projections and Drawing Conventions 1

1.1 Introduction 1
1.2 First-Angle Projection 1
1.3 Third-Angle Projection 3
1.4 Comparing First- and Third-Angle Projections 6
1.5 Non-Orthographic Projections 6
1.6 Patterns 8
1.7 Sections 10
1.8 Staggered Sections 11
1.9 Comments 14

Chapter 2 Basic Dimensioning Methods 15

2.1 Introduction 15
2.2 Creating 3D CAD Models 16
2.3 Chain, Parallel and Ordinate (or 'Running') Dimensions 18
2.4 Hole Dimensioning 19
2.5 Rotational Solids 23
2.6 Counterbores and Countersinks 26
2.7 Auxiliary and Dual Dimensioning 27
2.8 Dimensions Not-to-Scale 27
2.9 Arcs and Chords 28
2.10 Edges: Chamfers and Fillets 28
2.11 Curves and Surfaces 29
2.12 Thin Parts 31
2.13 Developed Parts 32
2.14 ISO Standard Methods and Symbols 32
2.15 Comment 33

Chapter 3 Introduction to Tolerancing 34

3.1 Introduction 34
3.2 Maximum Material Condition and Least Material Condition: MMC, MMS, MMR, LMC, LMS, LMR 40
3.3 Envelopes, Dependency and the Taylor Principle 41

3.4 Independence ... 42
3.5 Common Depictions of Tolerances ... 44
3.6 Limits and Fits with Grades and Classes ... 45
3.7 Using ISO 286-1:2010 and ISO 286-2:2010 ... 47
3.7.1 Hole-Based versus Shaft-Based ... 47
3.8 Examples ... 52
3.9 Comments ... 54

Chapter 4 Threads, Splines and Knurling ... 55

4.1 Introduction ... 55
4.2 Screw Threads ... 57
4.3 'M' Series ... 57
4.3.1 Tolerancing ... 61
4.3.2 Thread Engagement ... 65
4.3.3 Chirality or 'Handedness' ... 66
4.3.4 Multi-Start Threads ... 66
4.4 Unified Threads ... 67
4.5 Whitworth Threads ... 68
4.6 BA and Small Threads ... 69
4.7 Pipe Threads ... 69
4.8 Square, ISO Trapezoidal, Acme and Buttress Threads ... 70
4.8.1 Square Threads ... 70
4.8.2 Trapezoidal Threads ... 71
4.8.3 Acme Threads ... 72
4.8.4 Buttress Threads ... 72
4.9 Plastic Threads ... 73
4.10 Drafting Conventions ... 74
4.10.1 Male Threads ... 74
4.10.2 Female Threads ... 74
4.11 Splines ... 75
4.12 Knurling ... 76
4.13 Comments ... 77

Chapter 5 Introduction to Geometric Tolerancing and Form Tolerances ... 78

5.1 Key Concept—Features ... 78
5.1.1 Internal versus External Features ... 78
5.1.2 Features of Size ... 79
5.1.3 Integral versus Derived Features ... 79
5.1.4 Nominal, Real, Extracted and Associated Features ... 80
5.2 Maximum Material Virtual Condition (MMVC) and Least Material Virtual Condition (LMVC) ... 80

5.3 Geometric Methods of Dimensioning, Tolerancing and Validation ... 82
5.4 Form Tolerances ... 82
5.4.1 Straightness ... 83
5.4.2 Flatness ... 84
5.4.3 Roundness ... 85
5.4.4 Cylindricity ... 86
5.4.5 Profiles ... 86
5.5 Comments ... 89

Chapter 6 Datums ... 90

6.1 Datums and Datum Systems ... 90
6.1.1 Datum ... 91
6.1.2 Datum System ... 91
6.1.3 Datum Feature ... 91
6.1.4 Datum Simulator ... 91
6.1.5 Situation Feature ... 91
6.1.6 Display of Datums ... 93
6.2 Datum Systems ... 94
6.2.1 Degrees of Freedom ... 94
6.3 Problems with Common Validation Methods ... 98
6.3.1 How Flat Is a Surface (Flat) Plate? ... 98
6.4 Combined and Constrained Datums ... 100
6.5 Common and Combined Zones ... 100
6.6 Constrained Zones: Datum Targets ... 100
6.6.1 Example—Datum Targets on a Flat Plate ... 102
6.7 Contacting Features ... 102
6.8 Theoretically Exact Dimensions (TEDs) ... 103
6.8.1 Examples: Traditional and GPS Style Positional Tolerancing ... 103
6.9 References ... 105

Chapter 7 Orientation Tolerances ... 106

7.1 Introduction ... 106
7.2 Indicators Adjacent to the Tolerance Indicators ... 106
7.3 Parallelism Tolerance Indicator ... 108
7.3.1 Example—Plane Parallel to Planar Datum with Independence Principle ... 108
7.3.2 Example—Plane Parallel to Planar Datum with Envelope Principle Invoked ... 108
7.3.3 Example—Centreline Perpendicular to Plane ... 110
7.4 Perpendicularity Tolerance Indicator ... 110

7.4.1 Example—Centreline Perpendicular to Datum Plane A Parallel to Datum B 110
7.4.2 Example—Centreline Perpendicular to Datum Plane A 111
7.5 Angularity Tolerance Indicator 112
7.5.1 Example—Hole at an Angle to a Plane 112
7.5.2 Example—Integral Feature (Plane) Angled to a Datum Plane 113
7.6 Line Profile Tolerance Indicator 114
7.6.1 Example—Profiles with Form Tolerances and with Form and Orientation Tolerances 114
7.6.2 Example—Profile Tolerance Indicator with Adjacent Indicator 115
7.7 Surface Profile Tolerance Indicator 116
7.7.1 Example—Surface Profile of a Unified Feature 116
7.7.2 Example—Surface Profile of a United Feature with 'All Around' Symbol 117
7.7.3 Example—Surface Profile with a Combined (Tolerance) Zone 118
7.7.4 Example—Surface Profile with Unequal Tolerance Zone 119

Chapter 8 Location Tolerances 120

8.1 Introduction 120
8.2 Position Tolerance Indicator 120
8.2.1 Example—Spherical Feature 121
8.2.2 Example—Hole Position 121
8.2.3 Example—Angular Position 121
8.2.4 Example—Pattern 123
8.3 Concentricity Tolerance Indicator 123
8.3.1 Example—Concentricity 123
8.4 Coaxiality Tolerance Indicator 124
8.4.1 Example—Coaxiality 124
8.5 Symmetry Tolerance Indicator 124
8.5.1 Example—Symmetry of a Slot 125
8.6 Comments 125

Chapter 9 Run-Out Tolerances 127

9.1 Introduction 127
9.2 Run-Out 128
9.2.1 Example—Simple Run-Out 128
9.2.2 Example—Common Datum Run-Out 129

9.2.3 Example—Cone Run-Out 129
9.2.4 Example—Curve Run-Out 130
9.2.5 Example—Curve Run-Out with Direction Feature 131
9.2.6 Example—End Face Run-Out 131
9.3 Total Run-Out 132
9.3.1 Example—End Face Total Run-Out 132
9.3.2 Example—Total Run-Out on Cylindrical Surface 133
9.4 Comments 134

Chapter 10 Curves and Surfaces 135

10.1 Introduction 135
10.2 Basic Curve Geometry 136
10.3 Interpolation Splines 137
10.4 Approximation Splines 138
10.4.1 Bézier Curves 138
10.4.2 Continuity 140
10.4.3 Benefits of Bézier Curves 143
10.5 B-Splines 143
10.5.1 Example—Uniform Quadratic B-Spline 144
10.5.2 More Complex B-Splines 146
10.5.3 NURBS 147
10.6 Curves in Practice 147
10.6.1 Conics 147
10.6.2 Power Series Curves 148
10.6.3 NACA Profiles 148
10.7 Good Practice for Curve Profiles 150
10.8 Conveying Shape 151
10.8.1 Patterns 151
10.8.2 Coordinate Values 151
10.8.3 Curve Descriptions 151
10.8.4 Direct from CAD 151
10.9 Surfaces 152
10.10 Bézier Surface Patches 153
10.11 B-Spline Surface Patches 153
10.12 Rational Parametric Surfaces 153
10.13 NURBS Surfaces 153
10.14 Surfaces in Practice 153
10.14.1 Lofting 154
10.14.2 Sweeping 154
10.14.3 Bounded Surfaces 155
10.15 Continuity of Surfaces 156

10.15.1 Resurfacing 159
10.15.2 Blends 159
10.15.3 Curves Revisited 160
10.16 Comments 160

Chapter 11 3D CAD File Types and Transfers 162

11.1 Introduction 162
11.2 Basic CAD File Types 162
11.2.1 Neutral versus Vendor-Specific File Types 163
11.2.2 Text versus Binary 164
11.3 Translators 164
11.4 Common Data Types for 3D Product Information 165
11.4.1 CAD Vendor Specific 165
11.4.2 Standard Product-Level Formats 165
11.4.3 Kernel Systems 168
11.4.4 Mesh 169
11.5 Other Formats 171
11.6 Good Practice for File Saving/Exporting 171
11.7 Common Problems Opening Saved/ Exported Files 172
11.8 Comments 173

Chapter 12 Assemblies and Standard Parts 174

12.1 Introduction 174
12.2 Assembly Dimensions 174
12.3 Flexible Parts 174
12.4 Families of Assemblies 176
12.5 Standard Parts 177

Chapter 13 Metrology, Associative Features and Reverse Engineering 178

13.1 Introduction 178
13.2 Metrology and GPS 179
13.3 Direct Measurements 180
13.4 Indirect Measurements—Simulators 180
13.5 Derived Features 180
13.6 Scanning-Based Approaches 180
13.7 Meshes 182
13.7.1 Triangulations 182
13.8 Comments 184
13.9 Reference 184

Chapter 14 NC Code ... 185

14.1 Introduction ... 185
14.2 Example NC Program ... 185
14.3 File Types ... 186
14.4 Comment ... 187

Chapter 15 Additive Manufacture (AM) ... 188

15.1 Introduction ... 188
15.2 Why Use AM? ... 188
15.2.1 Machine Type Parts ... 188
15.2.2 Improved Part Designs ... 188
15.2.3 Generative Design ... 188
15.3 Additive Technologies ... 189
15.3.1 Binder Jetting ... 190
15.3.2 Directed Energy Deposition ... 190
15.3.3 Material Extrusion ... 190
15.3.4 Material Jetting ... 190
15.3.5 Powder Bed Fusion ... 190
15.3.6 Sheet Lamination ... 191
15.3.7 Vat Polymerization ... 191
15.4 Design for AM (DfAM) ... 191
15.4.1 Holes ... 192
15.4.2 Threads ... 193
15.4.3 Edges ... 193
15.4.4 Wall Thickness ... 193
15.4.5 Internal Structures ... 193
15.4.6 In-Process Shapes/Orientation and Part Stability ... 194
15.4.7 Unused Raw Material Removal and Disposal ... 194
15.5 File Types ... 194

Chapter 16 Surface Texture ... 196

16.1 Introduction ... 196
16.2 Profile versus Areal Methods ... 196
16.3 Profile Methods ... 197
16.4 Filtering ... 199
16.4.1 Amplitude Parameters ... 199
16.4.2 Spatial Parameters ... 204
16.4.3 Hybrid Parameters ... 204
16.4.4 Tolerance Acceptance Rules T_{max}, T16% and T_{med} ... 205
16.5 Areal Methods ... 205

16.6 Lay 206
16.7 Examples 207
16.7.1 Minimum Indications 207
16.7.2 Example 1—Minimum Requirements 207
16.7.3 Maximum Specification Options 208
16.7.4 Example 2—Full Specification 210
16.7.5 Example 3—Positions of Texture Indicators on Drawings 210
16.7.6 Example 3—Use of 'All-Around' Indicator 210
16.8 Comments 211

Chapter 17 Digital Product Definition, Drawings, 3D PMI and the Digital Twin 212

17.1 Introduction 212
17.2 Sheet Sizes 212
17.3 Line Types 213
17.4 Title Boxes 213
17.5 Simplified Drawings 214
17.6 3D PMI 214
17.7 The Digital Twin 216

Chapter 18 Welding 217

18.1 Introduction 217
18.2 System A and System B 217
18.3 Welding Processes 217
18.4 Welding Symbols 217
18.4.1 Square Butt 218
18.4.2 Single-V Butt 218
18.4.3 Single-V Butt with Broad Root 219
18.4.4 Single-Bevel Butt 219
18.4.5 Single-Bevel Butt with Broad Root 219
18.4.6 Single-U Butt 219
18.4.7 Single-J Butt 219
18.4.8 Flare V 220
18.4.9 Flare Bevel 220
18.4.10 Fillet 220
18.4.11 Plug 220
18.4.12 Resistance Spot 221
18.4.13 Projection 221
18.4.14 Fusion Spot 221
18.4.15 Resistance Seam 221
18.4.16 Fusion Seam 222

18.4.17 Stud ... 222
18.4.18 Steep-Flanked Single-V Butt ... 222
18.4.19 Steep-Flanked Single-Bevel Butt ... 222
18.4.20 Edge ... 223
18.4.21 Flanged Butt ... 223
18.4.22 Flanged Corner ... 223
18.4.23 Overlay ... 223
18.4.24 Stake ... 224
18.5 Supplementary Symbols ... 224
18.6 Example Elementary Weld ... 225
18.7 Example Combined Welds ... 225
18.8 Comments ... 225

Index ... 227

Preface

Anyone with some basic knowledge of 3D computer-aided design (3D CAD) modelling and an interest in communicating design information through engineering drawings, or by other printed or electronic means, will hopefully find this book useful, whether a student, researcher, professional or hobbyist. Central to efficient and robust communication is, of course, the use of a common language, but frustratingly, two competing systems that seek to establish such a language for product-based information are currently in operation, one controlled by the International Standards Organization (ISO) and one by the American Society of Mechanical Engineers (ASME). The two systems have much in common, but they differ, sometimes fundamentally, in critical areas. Given their complexity, it can be confusing to try to learn the two systems simultaneously, and, in the author's opinion, it makes a shorter learning path to start with one system and then learn the other, if necessary. For that reason, this book is focussed only on the ISO standards. Some comments and comparisons between the ASME and ISO systems are made for the reader to judge their relative merits, however.

The ISO standards cover many aspects of the specification of engineered products, including those that relate to terminology and symbols, as well as materials, manufacturing and metrology specifications. A core part of this involves methods for presenting the form, or shape, of manufactured products and any allowable variation. Most of us have an intuitive understanding of the concept of simple linear dimensioning and tolerancing as a means of specifying sizes applied to product shape, but more sophisticated geometric dimensioning and tolerancing (GD&T) methods offer a much richer means to control the variability that is inherent in any manufacturing-based project. The ISO Geometrical Product Specification (ISO GPS) offers a wide-ranging system of standards that define methods for achieving this.

Early chapters (1–4) of the book offer introductory coverage of some of the underlying traditional methods of presenting product form, dimensions and tolerancing that are still in current use and covered by ISO standards. Some readers, depending on their level of expertise, may skip or skim this material.

The remainder of the book is devoted largely to coverage of the ISO GPS system, including the depiction of product manufacturing information (PMI) and related topics, for example, computer numerical control (CNC) programming, additive manufacture, surface texture analysis and electronic transfer of 3D models.

It is hoped the book can be read sequentially in the order the chapters are presented or used as a reference that can be read in any order and dipped into when necessary. It is not exhaustive in coverage but should give the reader an introduction to many important topics that can then be followed up in other texts, especially standards documents and Internet resources.

Acknowledgements

I'd like to say thank you to Ed Tsang at the British Standards Institution for organizing my permission to reproduce material from the various BS ISO standards that I have referred to in this book. (Permission to reproduce extracts from British Standards is granted by BSI Standards Limited (BSI). No other use of this material is permitted.)

I'd also like to thank Dr James Young for allowing me to use some of the images and data from his undergraduate MEng project, which was completed at the University of Edinburgh.

At this stage it is normal for one to thank one's family, and I'd like to do so sincerely, although the book might have been finished much earlier if my grandchildren hadn't consistently insisted on climbing on top of me and demanding that I be a horsey or dragged me to play hide-and-seek or jump around on trampolines.

Author

Frank Mill is the Emeritus Professor of Digital Design at the University of Edinburgh, where he has taught manufacturing modelling subjects for over 40 years. He is a Chartered Engineer and Fellow of the Institution of Mechanical Engineers as well as a Principal Fellow of the Higher Education Academy (Advance HE). He holds a first-class honours degree in production engineering and a PhD in shape optimization. His teaching is based on his practical experience working as a support engineer for a leading CAD company as well as his design consultancy work for many engineering customers and his personal research in 3D modelling applications in manufacturing. He has taught undergraduate courses in engineering drawing, computer-aided design, manufacturing systems, digital manufacturing, product data management and engineering management as well as postgraduate courses in innovation and enterprise and in advanced 3D modelling.

Frank is a director of ShapeSpace Ltd, a company dedicated to providing software solutions for product modelling applications.

Introduction

3D Modelling and Downstream Information

0.1 GETTING STARTED: THE DIGITAL TWIN, INTERNET OF THINGS AND INTERNATIONAL STANDARDS

Those designers new to building 3D solid models will probably find an equal measure of fun and frustration as they develop their skills to create useful computer-based geometric models. Building these models may just be the start of the story; however, as the designer learns to generate all the additional associated data that may be needed for the total documentation that's necessary to support an engineering project. For the multitude of parties that may need access to some or all of this documentation, common means of communication are vital, hence the need for standards, whether by convention or more formally. It can be convenient to refer to the accumulated computer-based data relating to a product as its **digital twin**.

While there is no universally accepted definition of the digital twin, it can simply be thought of as a digital representation of the product that covers some or all aspects of its documentation for its administered lifecycle. Definitions of the concepts and language used in relation to the digital twin are given in the **ISO/IEC 30173:2023** standard. Figure 0.1 shows a relatively simple depiction of some of the documentation that might be involved in a design project; however, such information models will probably continue to grow in complexity as networked documentation has to cope with synchronization problems that result from design changes. In addition to all this, some product models are developed to support the **Internet of Things (IoT)**, whereby part models and associated systems are used to monitor and control products physically and remotely in the real world and in real time.

Central to the product development process is the need for a common language in which to depict products and product data, and to this end agencies such as the **American National Standards Institute (ANSI)** in partnership with the **American Society of Mechanical Engineers (ASME)**, **ISO** and other bodies such as industrial groups seek to develop standard ways to represent product information. Historically, this was largely achieved through engineering drawing standards, but now it includes those relating to other associated digital information, much of which is now supported by 3D computer-based models. Defining geometric constraints on product models has become an area where opportunities to improve existing crude methods of tolerancing have been exploited and

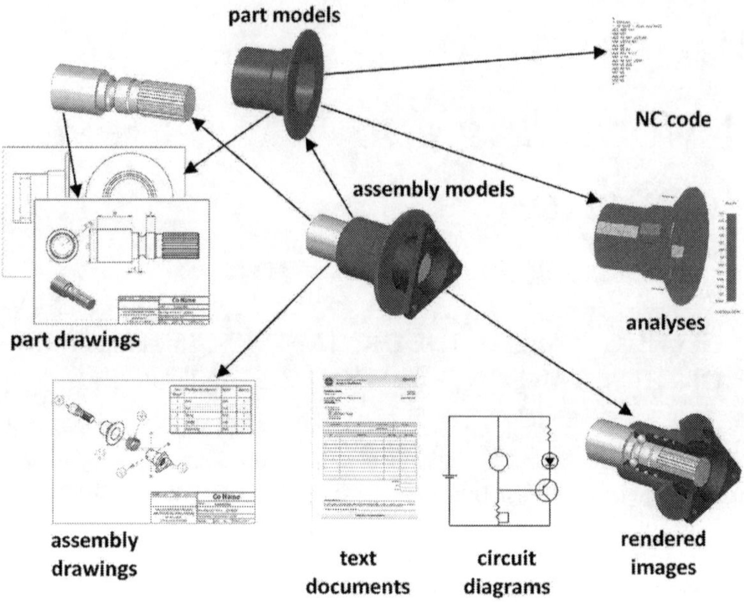

FIGURE 0.1 A depiction of a typical engineering product information environment.

largely standardized. At the time of writing, and for the foreseeable future, two systems of standards relating to geometrical tolerancing dominate the **ASME 14.5 Geometric Dimensioning and Tolerancing (GD&T)** standards and the **ISO Geometrical Product Specification (ISO GPS)** system of standards.

0.2 ISO AND ASME STANDARDS—KEY DIFFERENCES

Although simple systems based on ± tolerance values continue to be widely used, these can be ambiguous, and there is a growing body of opinion that tolerances which fully constrain the entire feature and part geometry should be used. Although more complex to depict in many situations, these types of tolerances can more clearly represent **design intent** and can lead to products that can be cheaper to make. This may simply be because they lead to a relaxation of manufacturing precision constraints, which is always likely to be welcome.

Both major standard systems are in wide use, but they are relatively complex to master. For this reason, this book covers the ISO system almost exclusively, as it can be difficult for those new to geometric tolerancing to learn two separate systems in parallel. Best to learn one first and then approach the other. There are many small details which differ between the ASME and ISO standards, particularly those that relate to individual geometric tolerance analysis, and these are beyond the current coverage of this book. The two systems of standards also differ fundamentally, in the author's view, in their most basic approaches and assumptions, which we should address.

The major differences in approach are as follows.

- The ASME standard has been developed with the philosophy that geometric constraints should be applied based predominantly on part function. The implication is that traditional measurement techniques (surface table, dial gauges and the like) might be used to validate the parts as they are made. While this is possible with the ISO system, in the latter there is more emphasis at the tolerancing stage on aligning with, and making use of, modern metrology methods such as those made possible by **laser scanning equipment** and **Coordinate Measurement Machines (CMMs)**.
- A second key difference between the two standards is that the ASME system is based on the default position that a simple tolerance should adhere to the **envelope requirement**. This means that if a shaft, say, is designated to be 10 mm ± 0,1 mm, then the finished part should be able to be measured along its length at opposite points, all of which should fall within tolerance AND the part should fit inside a perfectly cylindrical hole of 10,1 mm diameter. On the other hand, the ISO system works on the **principle of independence**, whereby all the measurements across the diameter should be within tolerance; however, such measurements are independent of form. In other words, a banana-shaped shaft could pass a tolerance test without it being able to fit within a perfect cylinder of 10,1 mm. Both systems have their advantages, as will be described in the chapters that follow.
- One further, more minor difference, but with potential for misunderstanding is that when using ASME values the '.' or period, acts as the decimal delimiter and commas act as the thousands delimiters, whereas in the ISO system a ',' or comma is used as the decimal delimiter with spaces as the thousands delimiters. Thus 1,000,1 mm (ASME) = 1 000,1 mm (ISO)!

If in doubt, check! is a common comment on engineering drawings.

0.3 THE ISO GPS STANDARDS

This book is based on the ISO GPS and related standards, of which there are many. To keep the entire standards system up to date, ISO has adopted a policy of publishing a wide range of individual standards, each of which can be updated readily or augmented with new standards. This is in contrast to the ASME 14.5:2018 standard, where the approach is to keep much of the material in a single document that will require a substantial upgrade from time to time.

The result of the ISO approach is that most of the standards refer to several other standards, largely through 'normative references', and so it can be difficult to interpret a standard from a single document at times. There are also occasions when a single standard is updated and the standards which reference it are out of date despite having a 'current' status. In the UK and those countries adopting **British Standards**, **BS 8888:2020** sought to bring together many of the BS and

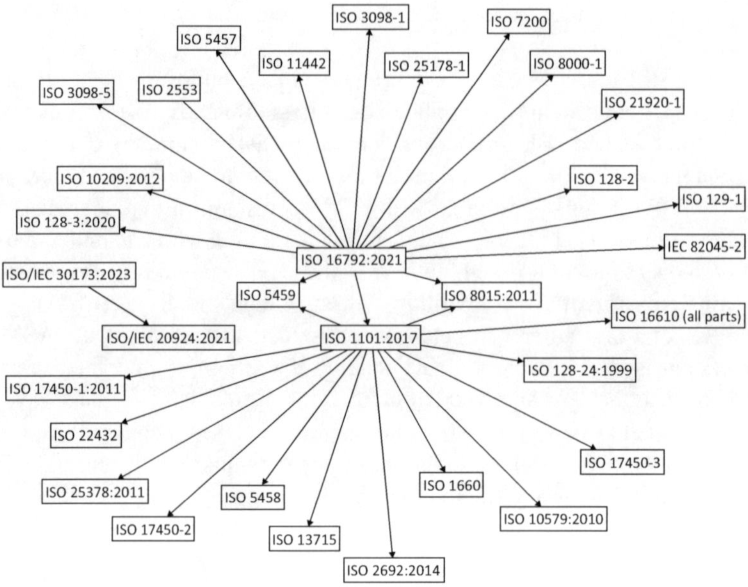

FIGURE 0.2 Partial network of ISO standards.

ISO standards, but it references several documents that have now been superseded with newer versions or withdrawn entirely. Standards relating to surface texture, for example, **BS EN ISO 4288** and several others, have been replaced with **BS EN ISO21920–1,2,3**. Figure 0.2 shows the main part of the local network of ISO standards that this book is based on. It is beyond most engineers to maintain a full, in-depth knowledge of all the standards that might need to be addressed in their day-to-day work, and this book does not attempt to cover them in entirety. Instead, the book is based on some of the overview standards with drilled-down detail as appropriate to give an overall understanding of the general principles of the ISO system. So, for example, if you are designing complex welds and need to add these to a drawing, hopefully, you can refer to this book to understand what this involves, and you may need to refer to **ISO 2553** for full detail.

Given the plethora of standards and their date-based definitions, it is important to show on engineering documentation what standards and what versions are in use.

0.4 ISO 16792:2021

The coverage in this book is largely based on the **ISO 16792:2021** standard, which points to many other standards but which gives an overview of the general ISO approach to '**Technical product documentation—digital product definition data practices**'. This covers how modern product definitions can include many forms of digital and paper-based documentation, for example, those referred to in Figure 0.1, and how these might be depicted in 3D documentation. The standard refers to many of the other high-level standards in specific areas of interest, for example, the major GPS standards and those for, say, welding or general drawing production or the specification and measurement of surface texture.

0.5 GOOD PRACTICE FOR 3D MODEL BUILDING

Before beginning any output creation process it is, of course, necessary to develop core 3D models in the form of CAD parts and assemblies. Although this book is not about 3D modelling itself, it is worth considering some general good practices that can be followed, which will help the reader build accurate and robust models—models that will duly support downstream applications. In this respect, it is generally useful for a designer to develop some level of understanding of the inherent limitations of 3D CAD systems, many of which are due to the challenges of discrete geometry.

System calculations may lead to results that involve point positions stored to many (17 or so) decimal places, and it is hoped that any rounding errors are contained at the extreme digits, while any tolerances of interest may occur much nearer the decimal point. Tolerances of interest may involve those that dictate when two values are 'equal' and hence when two-point positions are considered the same. Understanding this might help you to cope with mouse clicks that don't attach to points you'd like them to or that attach to positions you don't want them to. It might also explain why errors relating to non-closed profiles or non-watertight geometry occur and why models may be difficult to import from other systems, especially if you're trying to build solid models from scanned data. It also explains why many CAD systems struggle to cope with very small or very large sizes and why 'slivers' or triangles with high aspect ratios lead to trigonometric calculations that are troublesome. Nowhere are the problems inherent in approximating idealized or nominal points more pronounced than in the area of curve and surface modelling, where very small changes in the position of a point can lead to large changes to a curve profile. Difficulties can occur when trying to match curve/surface end point positions as well as the curve first and second (or higher) derivatives at joins. Chapter 10 describes some curve/surface CAD modelling in some more depth, but what follows here are a few rules that might help you build models that will work well and hopefully keep the levels of inevitable user frustration to manageable proportions.

0.6 PLANNING

Spend some thinking time before starting up your CAD package. You might like to think about how the geometry of the part(s) might be built and how this will support your downstream activities. In other words, ask what the purpose of the model is and what strategies will lead you to create robust 3D models that will support your design intent.

0.7 HISTORY-BASED OR DIRECT MODELLING?

CAD package permitting, you may want to decide whether to adopt a **history-based (parametric modelling)** or a **direct modelling** or even a hybrid modelling approach to your project. History-based methods allow you to interactively create a sequence of modelling steps with defined parameters that can be changed later. Changes can be made to geometric **features** created at an early stage, and new ones can be inserted. This requires some skill, but if done well will result in models that you can come back to later and readily understand and edit. Direct modelling involves the CAD user interacting with feature geometry by manipulating faces on features that can be pulled and rotated subject to geometric constraints the user can set at will. This also requires some skill, but it allows editing to be carried out without having to understand past design strategies and it also makes it easier to edit models regardless of the CAD system they were originally created in.

0.8 SKETCHING

When you're actually building your 3D solid models, you will want to make sure they are robust so that editing will be reasonably reliable and your model/CAD package won't crash too often. This is easier said than done, but in general, since most CAD packages allow 3D shapes to be built by projecting 2D shapes, good practice begins by creating good initial 2D sketches on xy, xz or yz planes or on user-defined planes related to these, for example, parallel or angled to them.

When creating 2D sketches, most CAD systems will automatically insert relationships between elements, for example, by adding end-connectedness between lines sequentially drawn or by making lines horizontal or vertical if these are drawn close to that condition. You will wish to add to these relationships by adding new ones, for example, 'equals' that will make sure two elements such as lines are always the same length, or you might add other geometric relationships like parallelism, perpendicularity and co-linearity. Using these will minimize the number of dimensioned sizes that are then needed to fully define your 2D shapes. Lastly, if any dimensions are to act as variables in later operations, it is worth taking the time to rename these to something meaningful so that you will recognize these subsequently.

0.9 3D OBJECTS

With robust 2D sketches on common reference planes, the geometric elements can be projected or copied onto profiles that are then extruded or rotated to form 3D features. If many of the model features are referencing the same profiles, they are more likely to update in a satisfactory synchronized way when you later make changes to the initial shapes. It is also often possible to add 3D relationships to create further geometry using geometric operators such as mirroring, although such commands are not always robust. Getting to know what operators are most reliable is quite CAD specific, and you will learn which ones you'll readily rely on and which ones to use with caution.

0.10 COMPLEX SURFACES

Good practice in building models is particularly important when creating complex shapes and surfaces, and it pays considerable dividends if the CAD user takes care to keep surface definitions as simple as possible. Many find creating complex surfaces particularly frustrating, especially when the CAD system doesn't seem to behave in the way they would wish. Building better curves and surfaces is helped enormously when the designer has some knowledge of the mathematical modelling techniques in use, and for that reason we devote an entire chapter (Chapter 10) to describing some of the common methods that are used to represent curves and surfaces. Regardless of this, it is generally useful to follow these very basic rules when building surface models.

- Use four-sided patches to create surfaces rather than triangular ones.
- Build larger patches than necessary when filling gaps in models and subsequently trim these to size.
- Do as much surfacing work as possible in a surfacing environment before attempting to change your surfaces to a solid thickened model.

0.11 FINISHING FEATURES

As you gain experience with your CAD system, you will learn what commands are most reliable. In general, for example, many CAD users like to add extra features like **chamfers** or **fillets** to models in 3D and towards the end of a modelling process rather than drawing these on the original 2D profiles. It can make the initial geometry creation easier and more robust, and late-modelled 3D finishing features can be readily suppressed to make simplified models if these are needed, for example, for finite element analysis or fast visualizations.

0.12 COMMENT

Once you are reasonably clear about what downstream outputs you may wish to create from your 3D models, you may wish to consider what standards these should adhere to and what problems you might have to overcome along the way.

Now is a good time to begin studying the practicalities and some of the theoretical aspects of making our outputs.

1 Basic Drawings
Common Projections and Drawing Conventions

1.1 INTRODUCTION

There are two viewing standards that are commonly used in engineering drawings: **first-angle projection** and **third-angle projection**. Both systems are in common use, and although each may be preferred in certain countries, you can come across either almost anywhere nowadays. It is therefore worthwhile to be familiar with both. To understand these viewing methods and why they are called what they are, we can begin by considering a simple set of orthogonal planes in 3D. Let's start with the first-angle method.

1.2 FIRST-ANGLE PROJECTION

The use of first-angle projection is widespread in many countries, particularly those in Europe and also in many parts of Africa, Oceania and South America. The label 'first angle' comes from the fact that it is based on imagining an object placed in the first quadrant of a set of Cartesian axes which are extruded to form orthogonal planes (see Figure 1.1).

The conical object shown in Figure 1.1 is imagined to be in the first quadrant with images from the top, front and sides projected past the object onto the planes behind, rather like shadows, or if the object is translucent they are similar to the projections we might get if a light with parallel rays was shone so that the outline and any edges could be seen on the planes behind.

A conical object has rotational symmetry around its central axis so we only need two views to fully describe the shape in 2D. We can imagine a definition of the first-angle projection in 2D by simply 'unfolding' two planes behind the cone as shown in Figure 1.2.

In fact this type of example forms the basis of the internationally accepted symbol that shows a drawing is in first angle. The symbol can be shown on an engineering drawing or its title box in either horizontal or vertical mode. The correct details for the horizontal depiction are shown in Figure 1.3.

We can of course extend this method of projection to any number of views of an object. Consider the part shown in Figure 1.4. The asymmetric nature of the object means that at least three views are needed to depict the object. In first-angle projection, all of these views are projected past the object onto an imaginary plane at the other side of the object and subsequently 'unfolded' in a manner similar to that used previously.

DOI: 10.1201/9781003395072-1

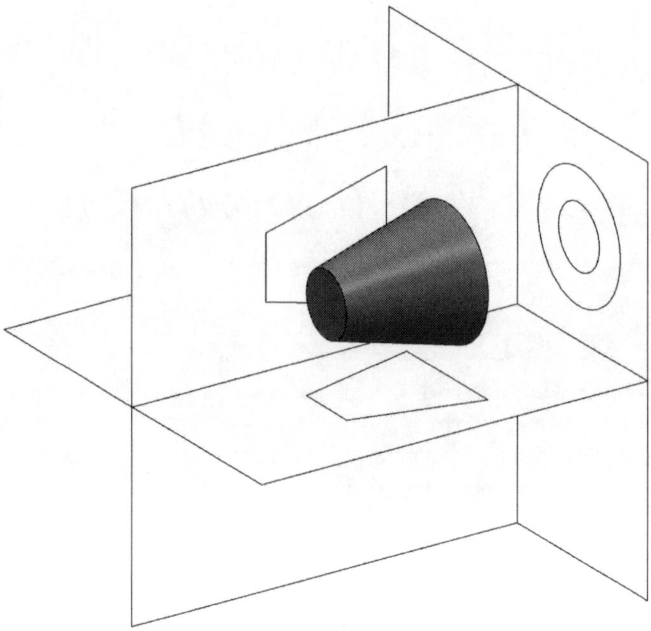

FIGURE 1.1 Orthogonal planes with a conical object shown in the first quadrant.

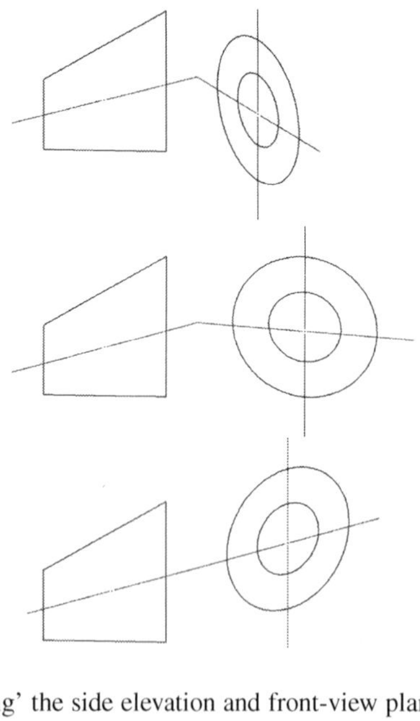

FIGURE 1.2 'Unfolding' the side elevation and front-view planes onto a 2D view in the first angle.

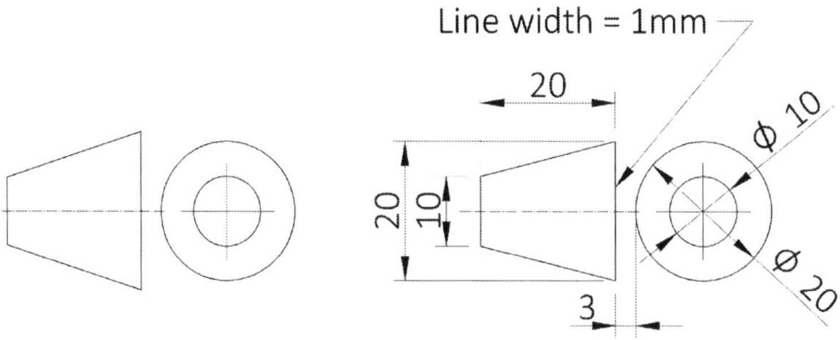

FIGURE 1.3 Symbol to be used on an engineering drawing using first-angle projection and suggested dimensions.

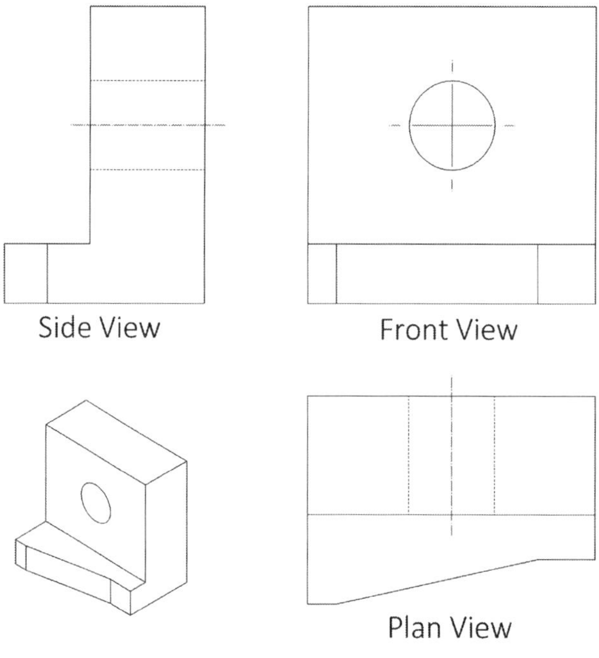

FIGURE 1.4 First-angle projection showing an asymmetric 3D part.

1.3 THIRD-ANGLE PROJECTION

In the US and some other parts of North America, third-angle projection is normally the preferred method used when creating orthographic engineering drawings. As you may already have guessed, the term 'third angle' comes from the fact that we imagine an object placed in the third quadrant of a set of planes created from Cartesian axes (see Figure 1.5).

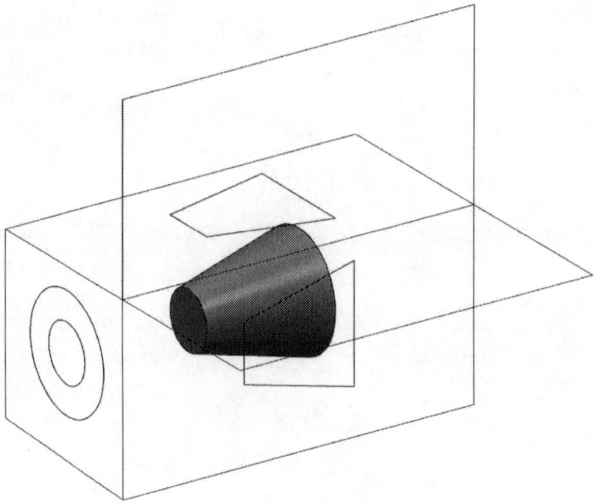

FIGURE 1.5 Orthogonal planes with a conical object shown in the third quadrant.

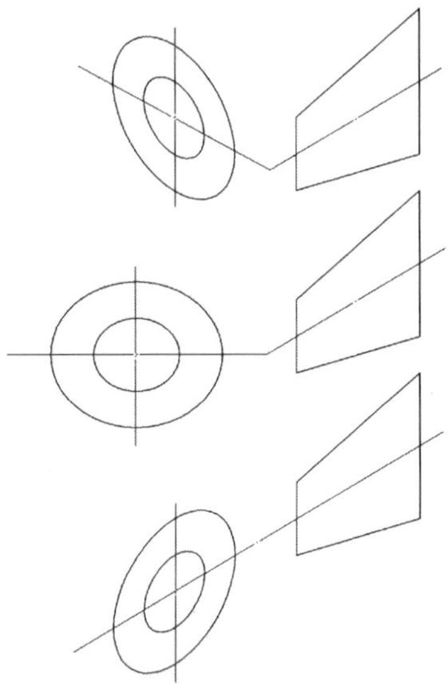

FIGURE 1.6 'Unfolding' the side elevation and front-view planes onto a 2D view in the third angle.

If we consider the side and plan views now, we observe the cone object *through* the orthogonal planes and so we project the image of the object onto these. The image will always be between the viewer and the object. When we unfold the front and side views we can obtain an image that can be used to depict third-angle projection as shown in Figure 1.6.

Thus, the following internationally recognized symbol for depicting the use of third-angle projection on engineering drawings is shown in Figure 1.7. Note that the dimensions can be scaled with line widths (e.g. 0,35 and 0,7).

Extending the third-angle method to a more complex object with no symmetry as before we can now create a simple engineering drawing as shown in Figure 1.8. In third-angle projection, all of these views are projected in front of the object onto an imaginary plane which is subsequently 'unfolded' along its common edges.

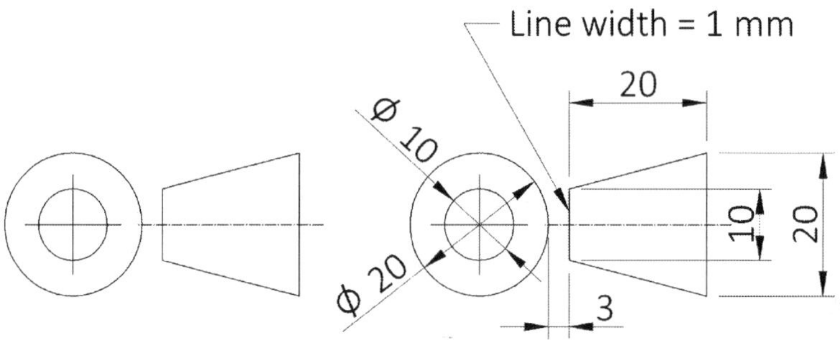

FIGURE 1.7 Symbol to be used on an engineering drawing using third-angle projection and suggested dimensions.

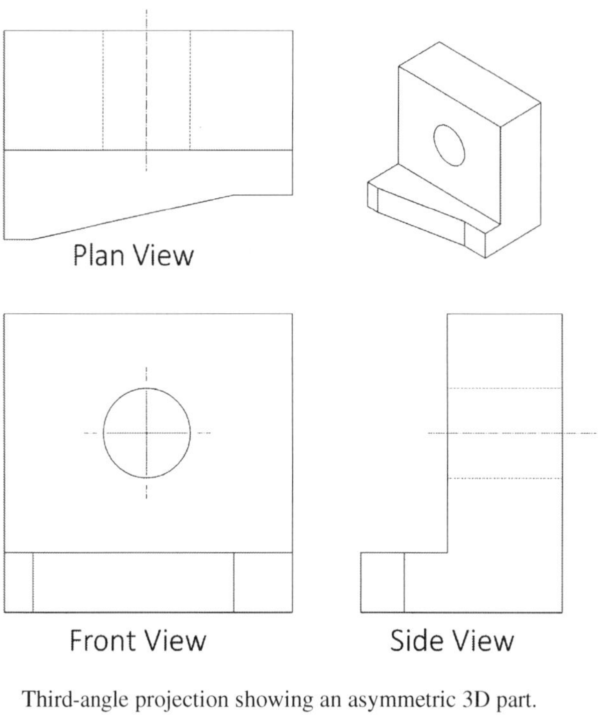

FIGURE 1.8 Third-angle projection showing an asymmetric 3D part.

1.4 COMPARING FIRST- AND THIRD-ANGLE PROJECTIONS

Both systems of projection show the same information, and so one method is not superior to another in this respect. The system to be used may be guided by the convention used in the particular country, company or local environment. In an increasing number of countries, notably in the UK and Canada, both systems are commonly used. Similarly in many countries both imperial and metric units of measurement might be used (more about that later).

For the beginner it is sometimes confusing when trying to remember the difference between first- and third-angle projections, and some find it useful to use the analogy of the Cartesian coordinate systems or simply to use the loose analogy that the first angle is like casting a shadow onto a back wall, whereas the third angle is similar to the cardboard cut-out toys that might be printed on cereal packets or in children's books.

If you're happy with the Cartesian analogy (even though it is not fully robust), you might like to consider why there is no use of second- or fourth-angle projection.

1.5 NON-ORTHOGRAPHIC PROJECTIONS

So far, we have considered only views that are at right angles to each other, but it is useful at times to add views that are derived from other viewpoints, particularly when looking normal to an angled face. As an example, observe Figure 1.9, which shows part of a toy.

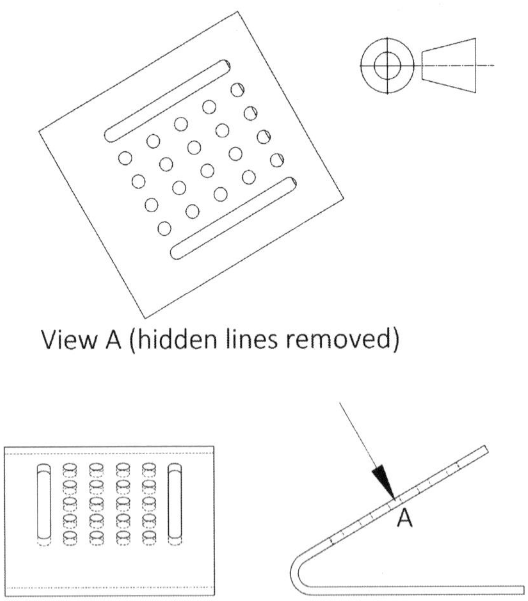

FIGURE 1.9 Object with angled face.

It may be best to use the view from A in the drawing to depict the pattern of holes and slots on the sloping face as it is most likely that we will want to dimension this view. Most people would find this the clearest view, and it is most likely to hold the dimensions that would be used by whoever is making the part. Of course, this is on the assumption that the purpose of the drawing is for someone to make the part from; it may also be for someone designing a complementary part in an assembly, or it might just form part of an instruction manual. The best way to start to create drawings is always to consider first the purpose of the drawing.

If we continue with the assumption that the drawing might be used by the object manufacturer, there are a number of improvements we can make that might help, even before we consider adding dimensions. For example, we might wish to make use of symmetry and create a drawing of only the portion that's considered necessary. See Figure 1.10.

If rotational symmetry is evident on the object to be made, we might make use of a **pitch circle diameter**, or **PCD** for short, and again we might reveal only a portion of the part, as shown in Figure 1.11.

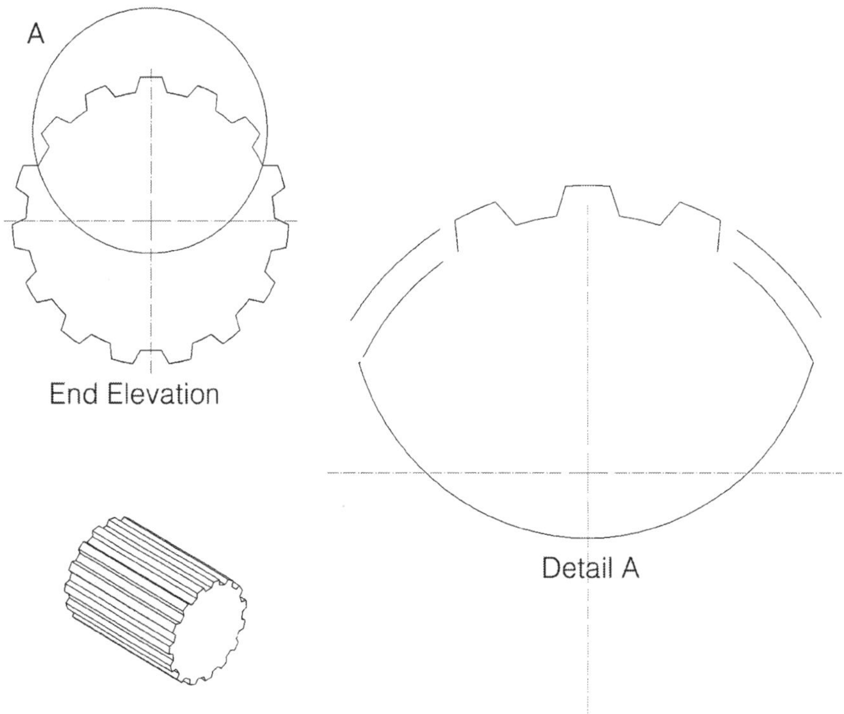

FIGURE 1.10 Symmetric component.

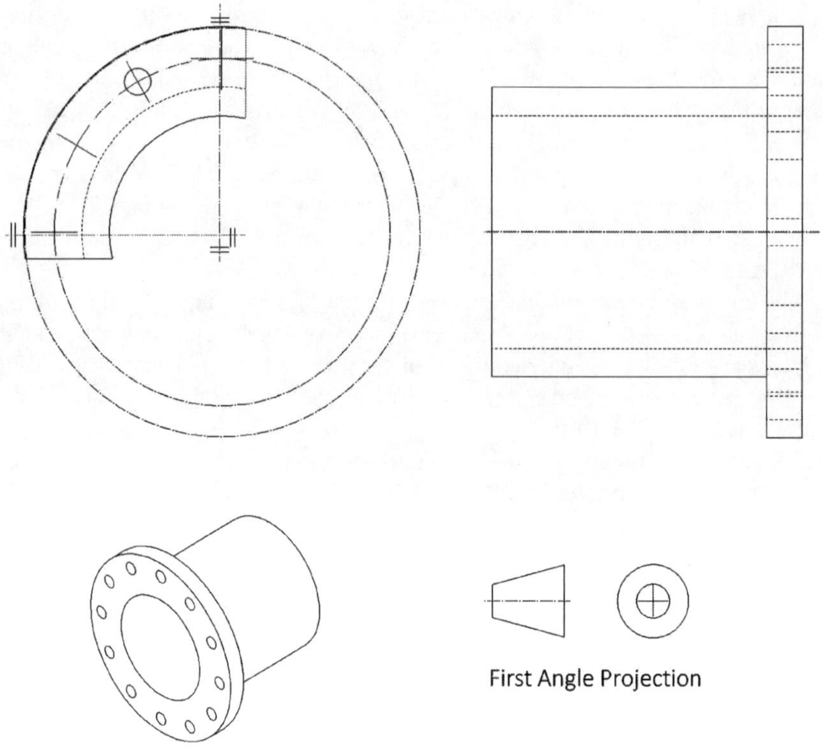

FIGURE 1.11 Holes around a PCD.

1.6 PATTERNS

Circular and rectangular patterns are available in most CAD systems and can save the designer considerable work in making patterned features such as gear teeth or hole grids. Traditionally, as is reflected in all international drawing standards, drawing many holes or teeth would be a tedious and unnecessary task, so shortcuts are widely used to simplify the drawing depictions. However, 3D users can create or show almost any geometric features that exist in 3D on a 2D drawing without any additional effort. It is as easy to show all 176 holes in the example shown in Figure 1.12 as it is to show one hole. The user must ask themselves, however, if it is clearer to include unnecessary detail. Again, we should consider the purpose of the drawing and who will be reading it.

Similarly, consider a spring that is used in a simple mechanism. By convention, the drawing of a spring can be simplified by showing only the first and last few turns, and this would commonly save considerable time in drawing the complete object with a pencil on a drawing board. From a 3D CAD model, however, it is often easier to show the whole spring, though perhaps not showing hidden detail.

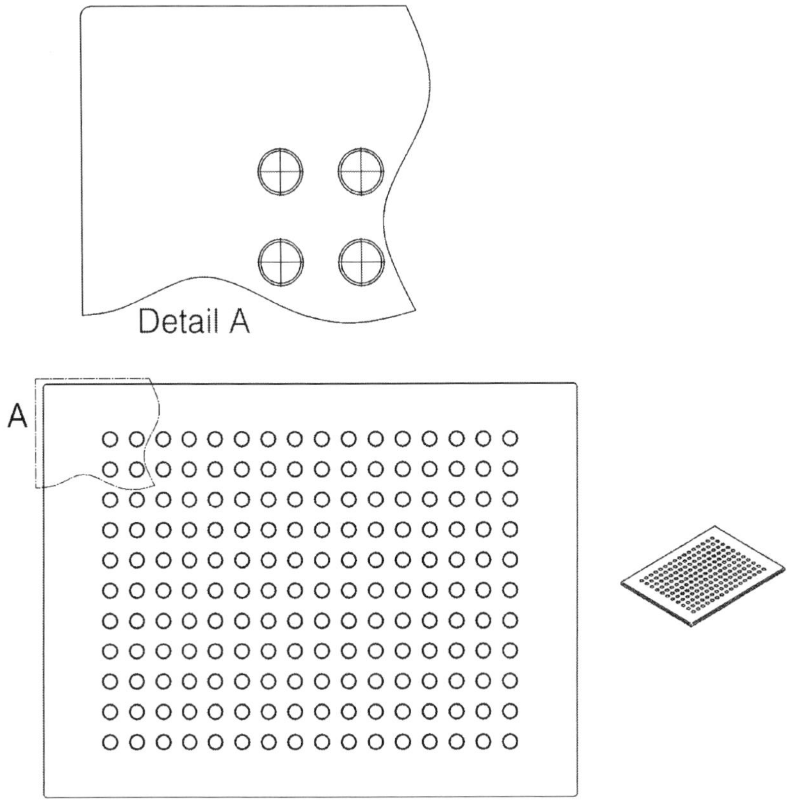

FIGURE 1.12 Platten with details shown.

As always, the draughtsperson must decide what is clearest for the person who is to receive the drawing. If it's an off-the-shelf standard spring that is to be bought, then there may be little point in a complex depiction; however, if it is to be manufactured, then complete details of the geometry, tolerances, material and manufacturing processes may need to be specified. Figure 1.13 shows several ways that a spring might be depicted in an engineering drawing. Note that it is also convention that **non-rigid** or **flexible** parts such as springs should be shown in their undeformed state. In assembly drawings, however, it may be necessary to show a flexible part in a deformed state.

Lastly, sometimes very **long parts** or parts that might come in various lengths can be shown simplified as **Broken Views**. Figure 1.14 shows two components, a long rectangular beam with a section removed in the middle so that only the end hole details are shown and a peg part showing how a cylindrical broken view is depicted. These views can be dimensioned in a number of ways depending on the purpose of the drawing.

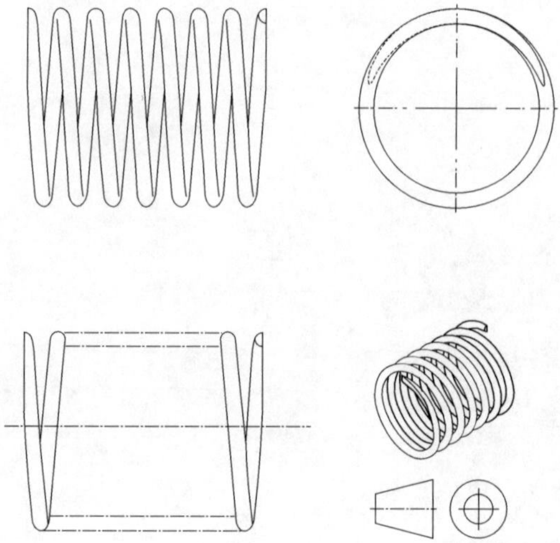

FIGURE 1.13 Spring depictions.

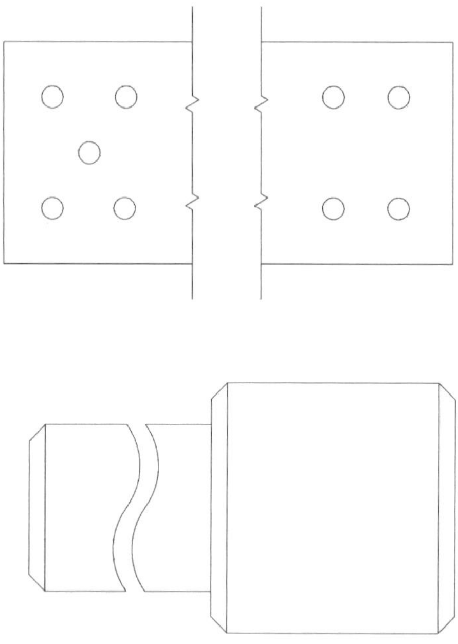

FIGURE 1.14 Broken views.

1.7 SECTIONS

With simple parts such as that shown in Figure 1.15, it is sometimes useful to show the object as a **section**. To do this we should show the cutting plane (shown as a line on an orthogonal view, e.g., the plan labelled A–A in the figure) and then use

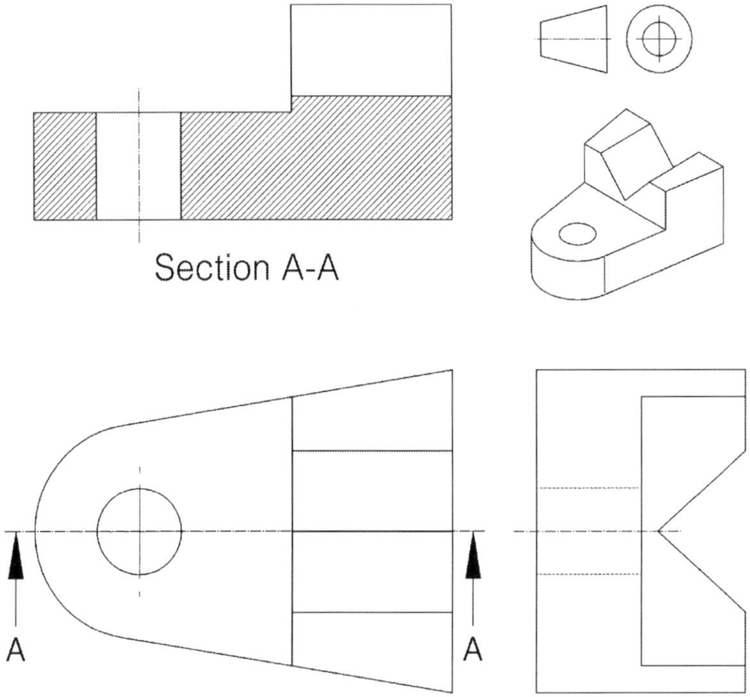

FIGURE 1.15 V-block showing section.

the relevant first- or third-angle convention to place a view that shows the resulting section view. The areas that have been cut away by the **cutting plane** are shown **hatched**, and most modern CAD systems can do this automatically.

There are more complex section views that can be created for parts; however, as 3D CAD systems can readily add 3D views, the need for sophisticated 2D sections is less than it used to be. They are sometimes still useful, however, and so it's worth considering some here. Since CAD systems generate much of this output automatically, it is useful to know how these views should look so that the correct command options are chosen and the user can validate the finished part.

1.8 STAGGERED SECTIONS

It's possible to create meaningful views of parts with more complex cutting sections that, for example, do not follow a simple plane but a staggered nonstraight profile line. Figure 1.16 shows such an example where a crank with an asymmetry is shown. The profile is swung into the vertical plane in the section view.

In Figure 1.17 a prismatic part is shown with the major feature cross-sections projected onto a side elevation view.

Before considering sections or cutaway views of assemblies, it is worth considering some of the basic guidelines for showing such views.

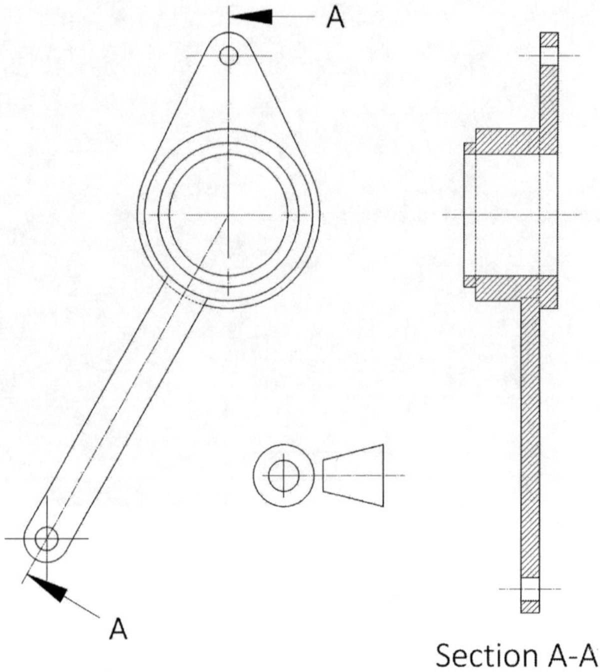

FIGURE 1.16 Crank arm.

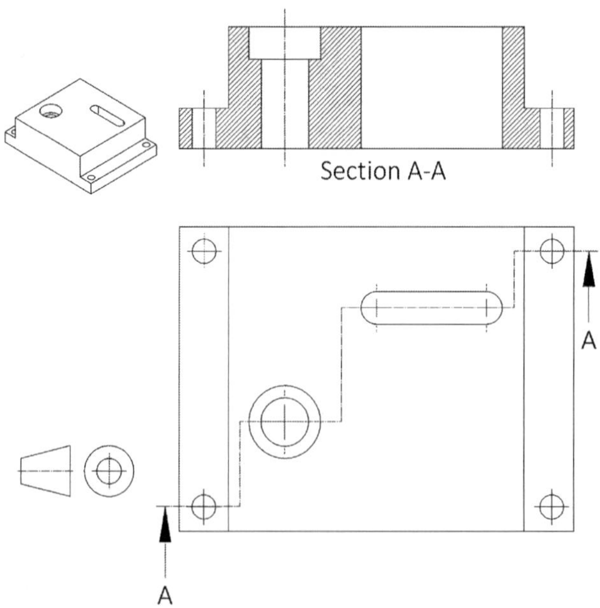

FIGURE 1.17 Sectioned block part.

Firstly, a simple view with only one sectioned area (see Figure 1.18) is usually shown with a hatching pattern at 45°, with the lines being spaced appropriately, normally more than 3 mm (or 1/8 in) apart, on an A2 or C wide drawing.

When cutaway areas on two parts are in close proximity or in contact we normally show the second part with the section pattern at right angles to the first, that is, at −45°. The lines on the two sections may have similar spacing, but their positions are staggered.

When a third part is added that's close or touches, it is possible to use 45° hatching again, but we would change the line spacing to differentiate the part. Most times this means decreasing the spacing but the user should choose according to the standard being used and the depiction that will give the clearest result to the person reading the drawing. Likewise, when we add more parts, we can change the hatching direction, angle, line spacing and even line type.

When assemblies are being sectioned or 3D cutaway views are created, it is normal to show cylindrical parts, for example, shafts in bearing assemblies, in their uncut non-hatched form. It is generally thought this is clearer.

Figure 1.19 shows a basic assembly with an example section view. Note that the spring part here is shown in a partly deformed state. See also Figure 1.13.

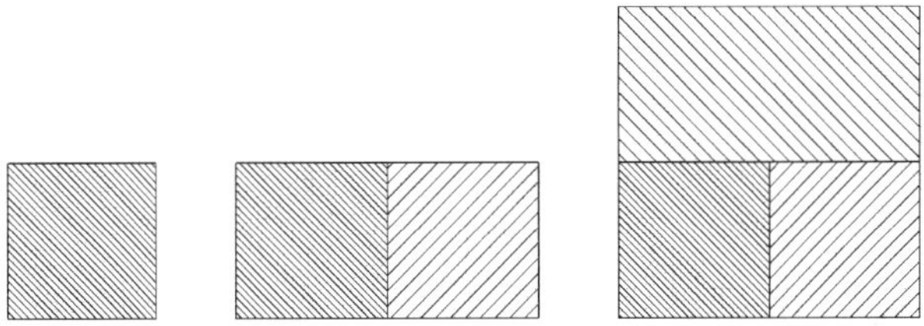

FIGURE 1.18 Hatching styles.

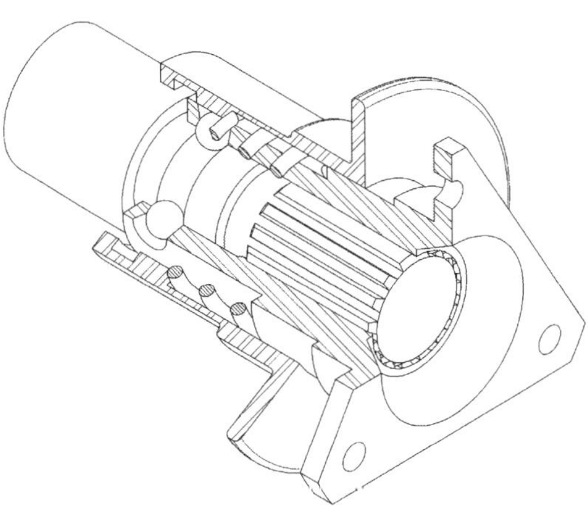

FIGURE 1.19 Quick disconnect.

1.9 COMMENTS

The methods discussed in this chapter represent some of the major types of views that are in widespread use on 2D drawings within the engineering industry. Most modern CAD systems will offer tools to make the generation of these view types simple and straightforward, and these will probably be based on standards that you can choose, typical ASME or ISO ones. For exact details the user can refer to individual standards documents; for example, further detailed orthographic projection rules are given in ISO 5456–21999. We will turn our attention now to basic methods of dimensioning engineering drawings.

2 Basic Dimensioning Methods

2.1 INTRODUCTION

The advice given in the following chapter is largely based on the needs of those producing engineering drawings for manufacture, but it's worth remembering that technical communications may also be produced for **user manuals**, **patent applications** and so on, and the consideration of the purpose of the drawing will influence the draftsperson's decisions.

Engineering drawings have evolved primarily to convey **design intent**, and they often consist of mainly **orthogonal views** of part outlines, together with **detail views** and **dimensions** depicting the various details of key parts. Traditionally, these were rarely complete definitions in a mathematical, or legal, sense, having many implied relationships; for example, concepts such as limiting values of **roughness**, **angularity** or **straightness** were often omitted from drawings but implied by expectations of the use of common sense in adhering to 'reasonable' values. This can work well where the designer and the manufacturer have a good mutual understanding of what 'reasonable' is, and it can make producing drawings cost efficient. In some environments a designer can talk to those in the manufacturing facility to clarify intent, either once a drawing is produced or, better still, beforehand. Sometimes prototypes can be produced to allow a design to be agreed upon or amended.

It is sometimes assumed that a designer knows how a part will be made, and this may be true in some circumstances, for example, where designers work at the same facility as production and within companies where designers may even have experience working in production. It is also common, however, that a part be designed prior to any investigation of manufacturing methods.

The aforementioned practices clearly carry an element of risk, the outcome of which may result in imperfect parts. This risk might be acceptable for parts that are cheap and/or made in small batches. However, in today's environment, it is increasingly the case that designers might have less-than-ideal contact and knowledge of the production processes. This may be a result of the modern specialized education and training, the rise of larger companies with discrete design and production facilities that are geographically remote on a national or inter-continental scale or the common use of outsourcing production to facilities in one or many plants located around the globe. Modern manufacturing methods also see the introduction of new processes being introduced at a rate that designers may struggle to stay aware of. All of this leads to the need for more formal and complete descriptions of design intent in all the aspects of documentation that may form the basis of legal agreements and contracts.

DOI: 10.1201/9781003395072-2

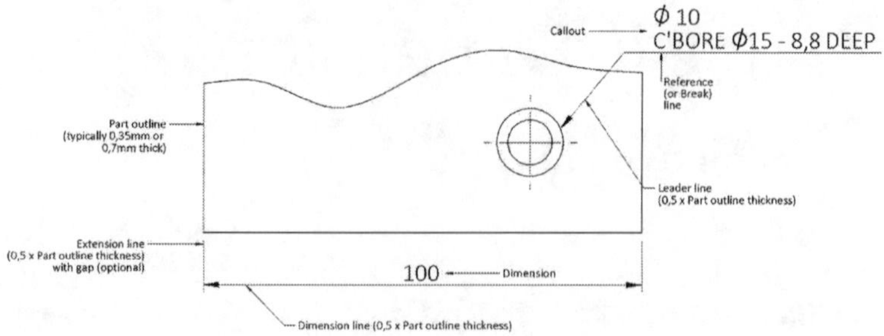

FIGURE 2.1 Elements of dimensioning.

What has all the above got to do with dimensioning? At one extreme the designer might be someone drawing up a simple plan for, say, a wedge-shaped door stop for an office, and the part will be made in-house. In such a case the designer will likely choose simple dimensioning methods. If, however, the designer is preparing drawings for a door wedge product that is to be made in batches of many thousands, which will be manufactured by a subcontracting company on the other side of the world, then it is highly likely that more robust methods are used to describe a design. These methods are the focus of later chapters of this book, and they will cover what is known as **Geometrical Product Specification** (or **GPS** as it's confusingly called in ISO parlance and not to be confused with the other GPS!). In the US, similar practices are referred to as **Geometrical Dimensioning and Tolerancing** or **GD&T**.

What follows in this chapter is a brief overview of commonly recommended dimensioning practices as described in **ISO 129–1:2019+A1:2021**. This explains methods for placing dimensions on 2D drawings and leads into basic methods of tolerancing dimensions covered in the next chapter. The methods described here are presented because most of them are included in the umbrella ISO standards, and those reading drawings are likely to come across them for a long time to come.

There are two broad styles of dimensions that are commonly used to describe features on a drawing, dimensions and **callouts**. A simple example of these and the nomenclature of the lines they relate to are shown in Figure 2.1.

2.2 CREATING 3D CAD MODELS

As discussed in previous chapters, if you're creating drawings or other dimensioned outputs from 3D CAD models, it's worth thinking about your communications early. Like many rules, these can be broken when the need arises, but it is generally a good idea to plan for creating drawings, **product manufacturing information (PMI)** documents or manuals, for example, that comply with agreed standards.

Crucially, consider whether, or how much, you can take into account how the part will be manufactured. There are often trade-offs between functionality and manufacture, and it is worth investing some thought on this upfront whenever possible. Modern CAD systems can make it easy to create complex geometry quickly, and it's generally a good idea when building a 3D model to keep control of your work by making your dimensioning clear and simple wherever possible.

Let's start with a trivial example, a basic door wedge, as shown in Figure 2.2.

Before producing any documentation of the part, it is worth considering what views are necessary and how you might go about documenting and dimensioning them. The entire part needs only three dimensions to fully define its shape, but which three should be used on the drawing? Hopefully the question would already be answered before the part was modelled and the important dimensions were chosen so that they might be taken out on a drawing with ease. Sometimes the dimensions that were considered important to meet design requirements may differ from those that are ideal with which to manufacture and validate the part. The designer may have to consider this and make whatever adjustments are acceptable.

The wedge may have been modelled using the base length, base width and slope angle. This is reasonable from a design point of view but might not be ideal for the manufacturer who might prefer to measure simple lengths rather than angles.

In the case of a door wedge, you can imagine that it so happens it was designed with the base length (135 mm), base width (40 mm) and height (40 mm) because the designer thought that this would meet their requirements and be easy to measure out and cut.

With modern CAD systems it is easy to take the 3D model onto a drawing view and simply click on some edges to insert dimensions, but this is often too easy. For example, it would be relatively straightforward to dimension the slope length (which works out in this case to be 140,801 mm)! Would we really want to tell our manufacturer this size? It implies a precision that is probably not required as it's unlikely a door wedge would need to be made to 1/1000th of a mm accuracy.

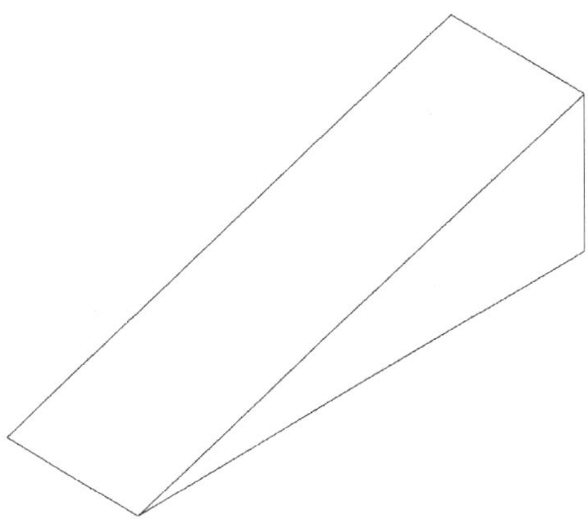

FIGURE 2.2 Simple door wedge.

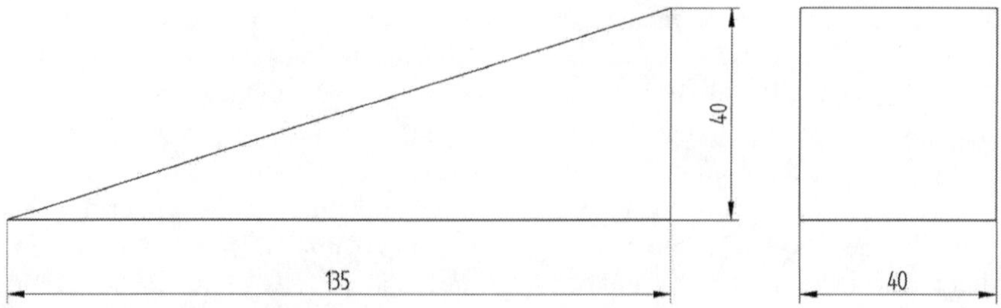

FIGURE 2.3 Simple-dimensioned orthogonal views.

This example may seem obvious, but it is a surprisingly easy trap to fall into when taking dimensions from an existing 3D model. Be warned!

So, we might produce a simple drawing like the one shown in Figure 2.3.

Such a simple diagram lacks a considerable amount of manufacturing information, for example, material or the allowable tolerances on the form or allowable surface roughness; we'll come to these later in the book. Similarly, it is unlikely that unfinished edges would be desirable so we should state what conditions these should be in. Again, more of that later, but for now we'll consider how the basic dimension numbers might be added in common situations.

Most CAD systems will have automatic support for the common international standards, but it is worth checking the settings on these to make sure they are in keeping with the standard (and version/year) with which you wish to adhere, for example, **ISO 129–1:2019+A1:2021**. Major points to note so far are that extension lines and dimension lines are narrower (usually half the width) of the part outline, which is normally 0,7 mm for A0, A1 or A2 sheet sizes and 0,5 mm for A3 and A4 sheet sizes. The extension lines may touch the part outline, but a small gap separating them from the part outline is also allowed by modern standards and is also a common practice. The dimension values are usually shown in the centre of and separate from the dimension lines and can be read from the bottom or right-hand side of the drawing sheet. Extension lines should ideally not cross each other for reasons of clarity, though this is often unavoidable, and the dimensions should, whenever possible, be outside the profile of the part.

2.3 CHAIN, PARALLEL AND ORDINATE (OR 'RUNNING') DIMENSIONS

Figure 2.4 shows three single views of a stepped block with alternative styles of placing dimensions. These are (a) **chain**, (b) **parallel** and (c) **running dimensions**. An engineering drawing might adopt any of these or even mix the styles depending on the purpose of each dimension and its associated tolerances.

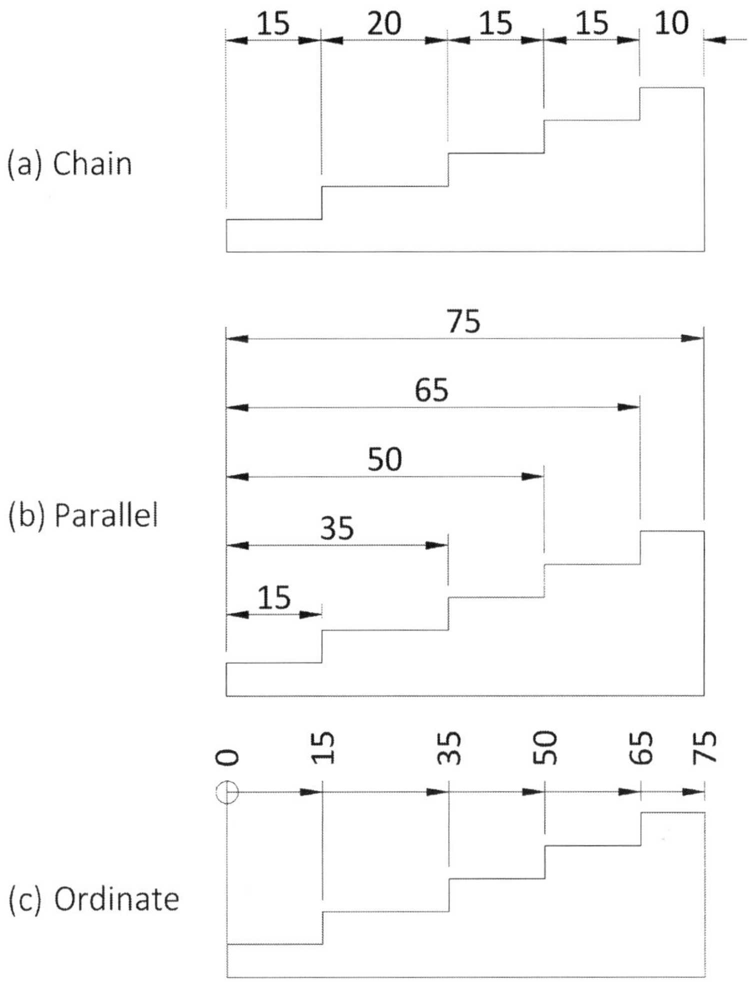

FIGURE 2.4 Basic dimension styles.

2.4 HOLE DIMENSIONING

If you are developing drawings for a workshop and have no issues in meeting design functionality requirements, the main consideration may be to keep manufacturing as simple as possible. If the part is to be made on a **computer numerical control (CNC)** machine, for example, it may be desirable to adopt chain dimensions if the CNC programmer is working on point-to-point machining or to use parallel or ordinate dimensions if the programmer is working in absolute mode (see Chapter 14 for further detail on numerical control (NC) programming). We can use ordinate dimensions in two directions, for example, X and Y to create **coordinate dimensions**. This is common practice when dimensioning groups of features such as holes. For example, see Figure 2.5.

However, undesirable tolerance build-up can occur with bad choices of dimensioning methods so the design intent means the designer will use those dimensions which ensure satisfactory design functionality.

Note also that the layout in Figure 2.5 has all hole locations dimensioned outside the part profile; however, all the hole-size dimensions are shown inside the hole spaces. This is necessary to keep the drawing relatively clear and to avoid placing dimensions within the part geometry. Common alternative methods of dimensioning holes are shown in Figure 2.6.

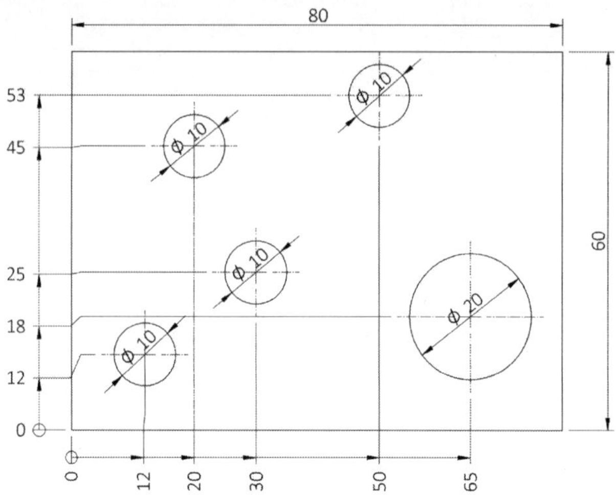

FIGURE 2.5 Coordinate hole positions.

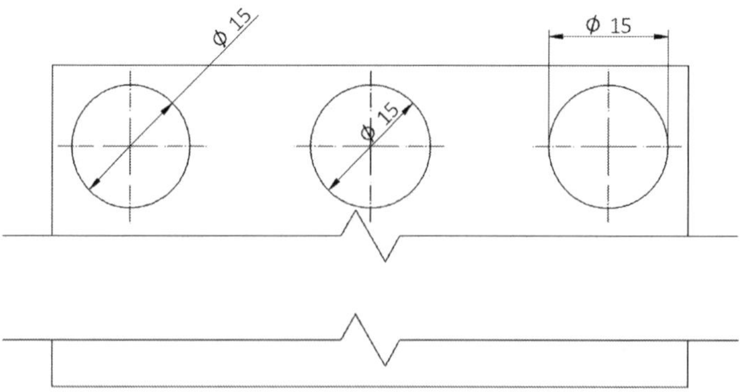

FIGURE 2.6 Sample hole dimensioning styles.

Note also that the part is shown with two broken lines indicating that it has been truncated to save space. Where a number of holes are of the same specification, it is also common practice to group these with a single comment. In most CAD systems this is achieved using a **callout** command. Figure 2.7 shows such a method for six through-holes with a 10- and 15-mm diameter counterbore to a depth of 8,8 mm using standard symbols rather than standard shortened names (as in Figure 2.1).

A further method that can be used to dimension holes is to make use of a **hole table** that can show in each row any desired properties of each individual hole, for example, its coordinates, diameter, tolerance or thread information. Figure 2.8 shows a simple hole table indicating the position and diameter of the previous five-hole layout (shown in Figure 2.5).

Note that when working with large groups of holes on a CNC machine, it is common practice in many workshops to use hole-drilling macros or subroutines. These often assume point-to-point machining so it may be worth checking whether you want to represent the hole position dimensions in point-to-point or absolute formats.

Note also that dimensions for hole diameters can be shown on **cutaway** and **partial views** as shown in Figure 2.9.

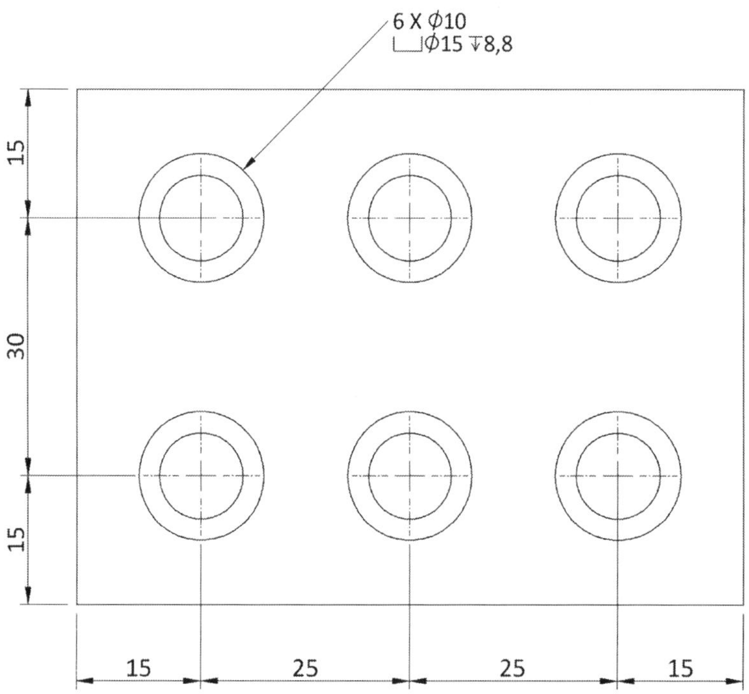

FIGURE 2.7 Counterbored hole pattern callout.

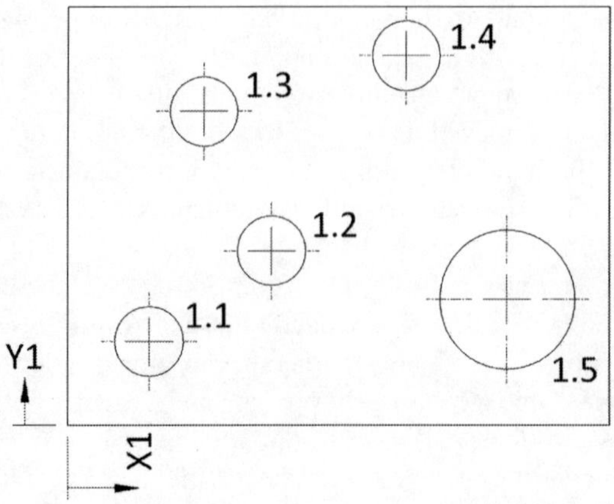

Hole	X	Y	Size
1.1	12 mm	12 mm	10 mm
1.2	30 mm	25 mm	10 mm
1.3	20 mm	45 mm	10 mm
1.4	50 mm	53 mm	10 mm
1.5	65 mm	18 mm	20 mm

FIGURE 2.8 Coordinate hole table.

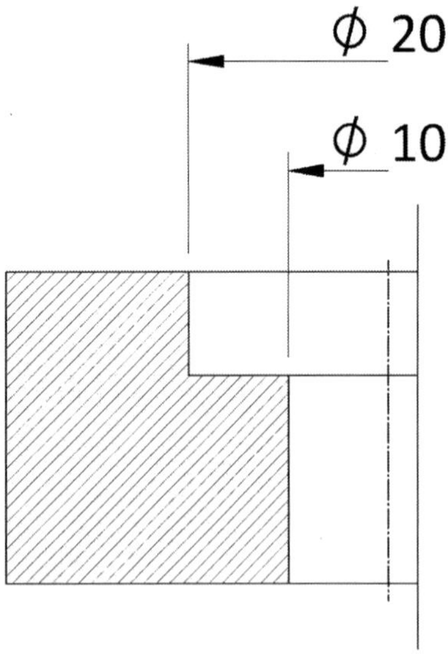

FIGURE 2.9 Diameters on partial views.

2.5 ROTATIONAL SOLIDS

Similar methods of dimensioning can be adopted when drawing rotational parts such as shafts. For example, Figure 2.10 shows a simple shaft with diameters shown in two different styles: with half-dimension lines and full-dimension lines. Note also that there is a taper on the shaft. This is dimensioned with the length (20 mm), the smallest diameter (30 mm) and the angle of the taper (30°). This implies that these are the measurements that should be made to verify that the finished part falls within any stated tolerances. Alternatively, the taper could be represented by specifying the base, height and length made by the 2D view, implying that these are the sizes that should be measured. Where design requirements, manufacturing methods and metrology regimes cannot be aligned, then the designer may need to calculate and show on a drawing such dimensions as are necessary to ensure that all demands can be met. In the previous example this might simply mean showing the taper's larger diameter, if this is easier to make/measure from, rather than the angle made. This will also mean computing suitable tolerance zones (see Chapter 3 onwards) for any sizes needed for verification purposes.

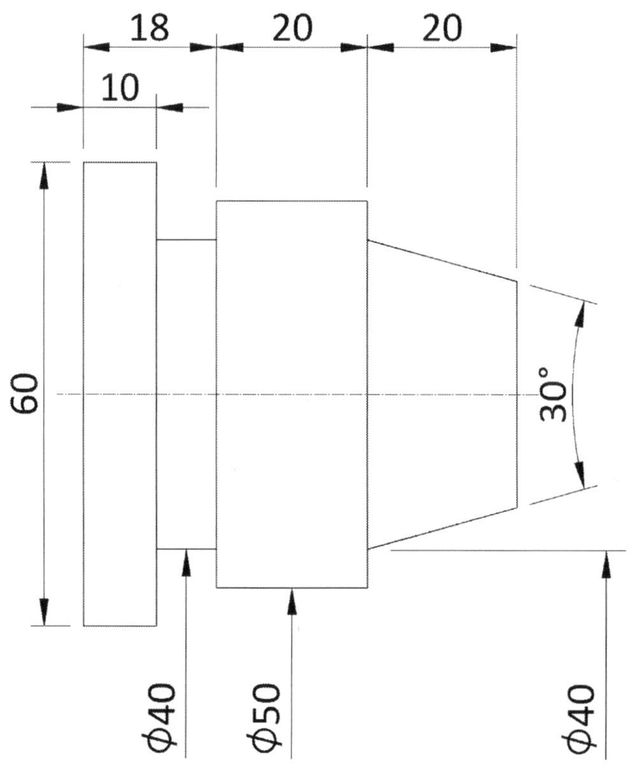

FIGURE 2.10 Stepped shaft.

To show that a diameter relates to a spherical surface, the letter 'S' is placed in front of the diameter symbol, as shown in Figure 2.11.

Spherical radii can also be depicted in a similar way, that is, by placing the letters 'SR' in front of the size. This is especially useful on surfaces where the spherical symmetry is not clear on the drawing, for example, see Figure 2.12.

Other features that commonly appear on shafts are those that are non-rotational, for example, **flats** and **keyways**. A typical flat area may be required so that torque can be applied via a spanner or for a location feature such as a square hole. Figure 2.13 shows a shaft with flat surfaces machined into it. The flat areas are shown with crossed lines and dimensioned across the flats; in the case of a square section, only one dimension is needed.

An alternative to using large flat surfaces is to use slots and keyways to apply (modest) torque to a shaft. Figure 2.14 shows two types of common keyways (side milled and end milled); there are others such as tapered keyways. Keyway assemblies are a convenient way to allow torque to be applied between the parts

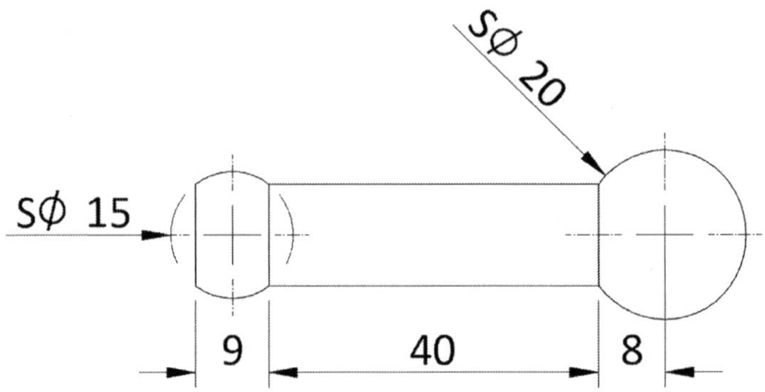

FIGURE 2.11 Spheres on the shaft.

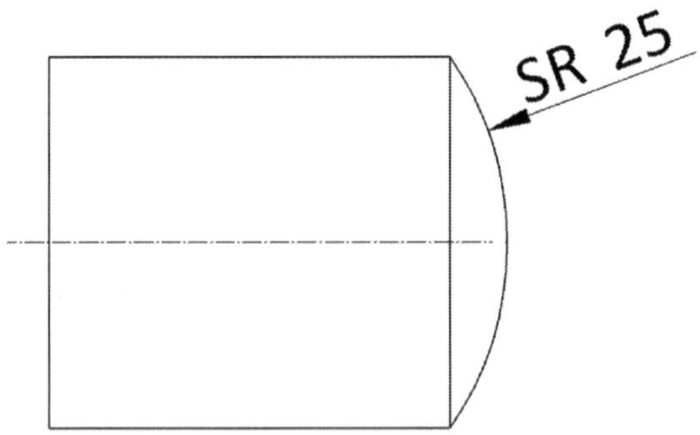

FIGURE 2.12 Spherical radius on the shaft.

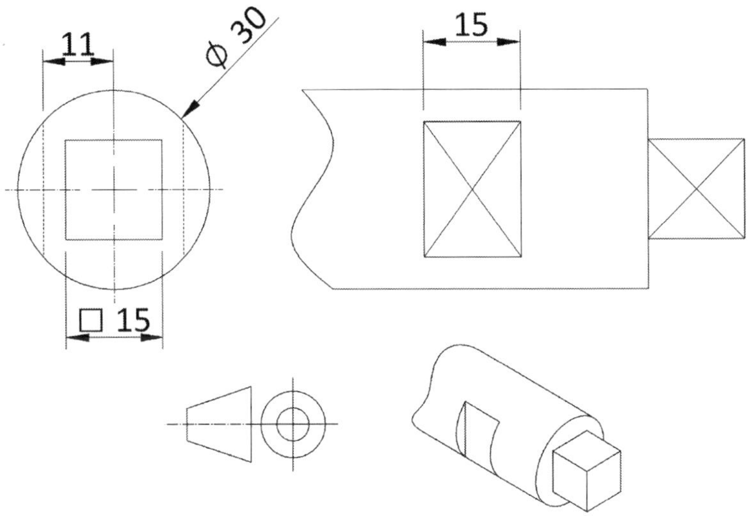

FIGURE 2.13 Shaft with flat surfaces.

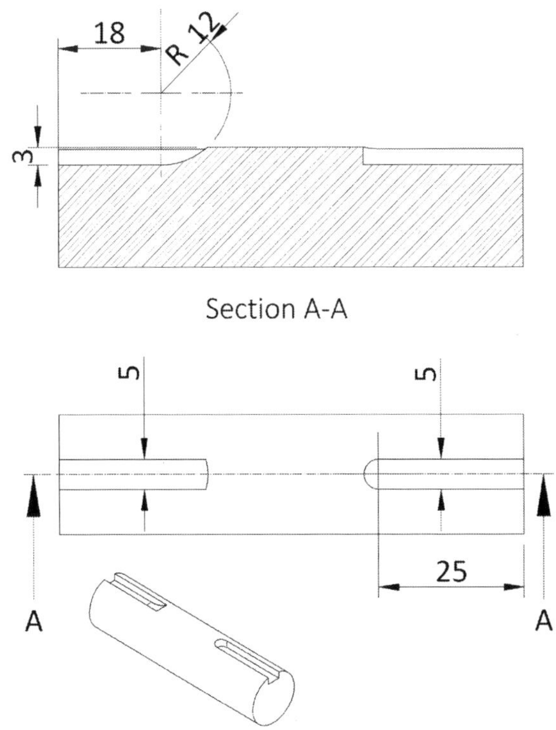

FIGURE 2.14 Keyways on the shaft.

of rotational assemblies because it would be difficult to machine protrusions in shafts or holes that might be used to engage the parts of an assembly. So, instead, a slot is machined in both the shaft and the hole, and a standard key is inserted between the slots to limit rotational motion by the shaft relative to the hole. The keys themselves can be made or bought as standard parts.

2.6 COUNTERBORES AND COUNTERSINKS

Previously an example of a **counterbore** was shown (Figure 2.7) using standard symbols on a callout to indicate the various diameters and depths. Alternatively, these sizes can be shown using simple dimensions as in Figure 2.15. However, it should be noted that common sizes of counterbores can be used, and the exact dimensions used should be based on those specified for a particular bolt. Counterbores are also commonly created using a counterbore tool so it can be worth checking available or preferred sizes.

Similarly, **countersinks** can be dimensioned using simple dimensions and/or standard symbols to show the large diameter of the countersink, as shown in Figure 2.16. Again, it is worth noting that dimensions should be checked against the sizes of the screws being used. The figure shows sizes for a typical screw head, but it may not be worth quoting diameters to 1/100th of a millimetre as it may imply a requirement for a measurement to be made. Also, it is important to show the angle of the countersink to ensure that a screw will bed in properly and the sloping face will bear any stresses evenly. Home woodworkers commonly use the end of a large drill to create a countersink, and although this can sometimes work for soft woods this is not good practice. The drill tip angle is unlikely to be the same as the countersink angle on a screw head.

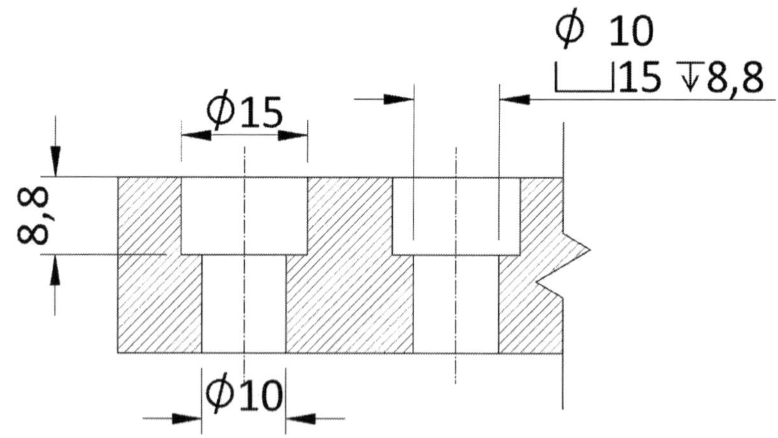

FIGURE 2.15 Alternative depictions of counterbored holes.

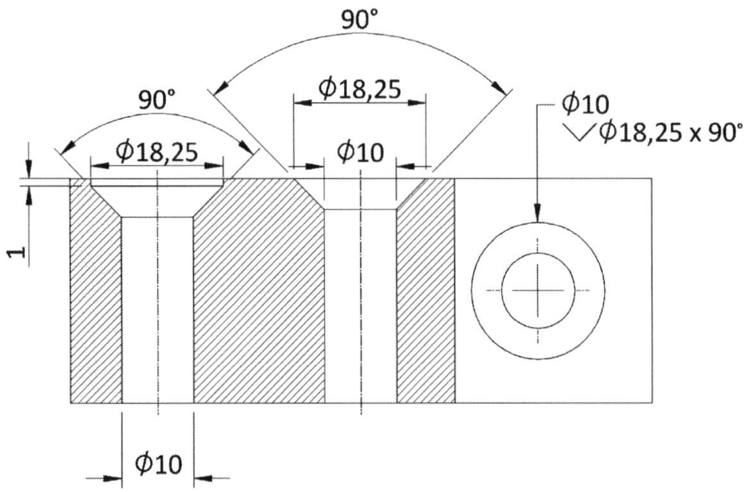

FIGURE 2.16 Countersink dimensioning.

2.7 AUXILIARY AND DUAL DIMENSIONING

Auxiliary dimensions are those that are used for illustration purposes only, and they do not carry tolerances. For example, in Figure 2.17, the part length is fully dimensioned by the hole positions. The overall length (100) is there only for illustrative purposes and should not have tolerances applied nor should the length be formally measured.

Sometimes dimensions can be required that show both **metric** and **imperial units**. This is common where parts have been designed, for example, in imperial units but may need a metric representation for manufacturing purposes, and it is a common problem in countries that formerly used imperial units but have switched to metric. It is also common for parts designed using metric units to fit into assemblies that were designed in imperial units. Representation of **secondary units** is normally achieved by showing them as auxiliary dimensions alongside or under the primary dimension. In the context of an environment based on metric units, the imperial units would be shown in such a manner. These should be placed in brackets and can be underlined. It is also sometimes convenient to show the secondary units on such dimensions. Figure 2.18 shows such dual dimensioning.

2.8 DIMENSIONS NOT-TO-SCALE

It is sometimes convenient to show dimensions that are shown on parts that are drawn **not-to-scale**, particularly where drawings are created as part of secondary or illustration-only documents. Figure 2.19 shows a dimension of 100 mm drawn on a not-to-scale figure, and the dimensional value is shown underlined.

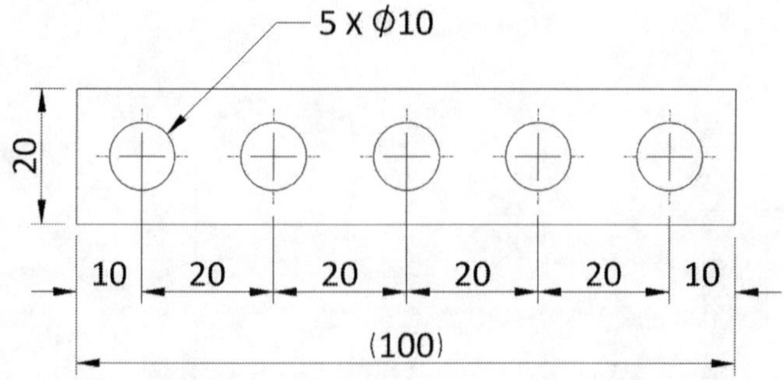

FIGURE 2.17 Auxiliary dimension (bracketed).

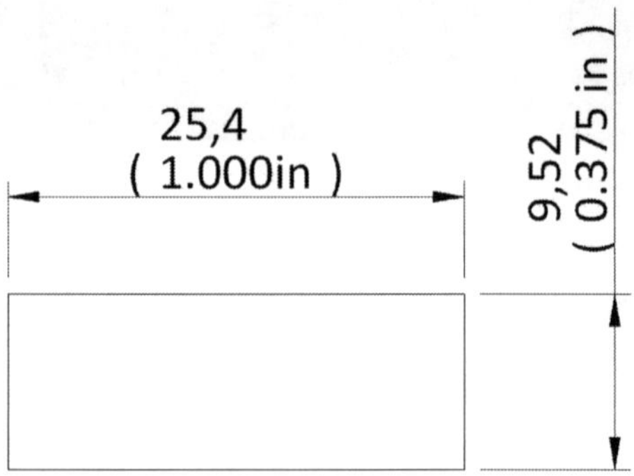

FIGURE 2.18 Standard and secondary dimensioning (metric and imperial).

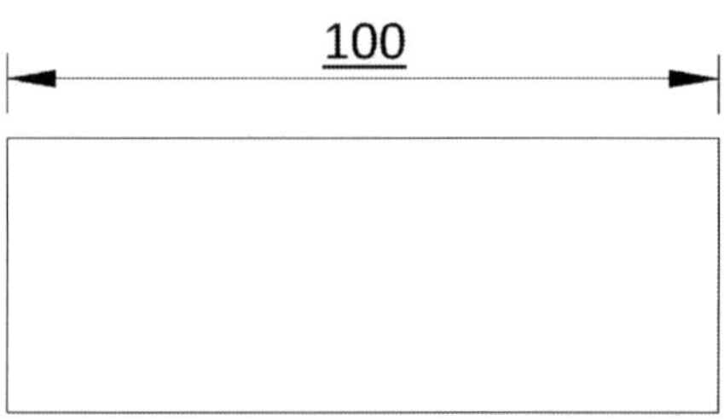

FIGURE 2.19 Dimension not to scale (therefore underlined).

2.9 ARCS AND CHORDS

Figure 2.20 shows two simple ways to depict the lengths on an arc, that is, by **arc length** (left) and by **chord length** (right).

2.10 EDGES: CHAMFERS AND FILLETS

There are several ways to depict edge conditions on drawings. Figure 2.21 shows some common ways to depict edge conditions such as **chamfers** and **fillets**.

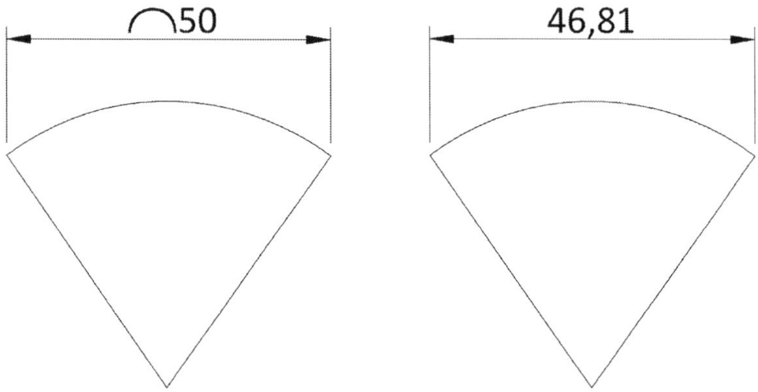

FIGURE 2.20 Arc length (left) and chord length (right).

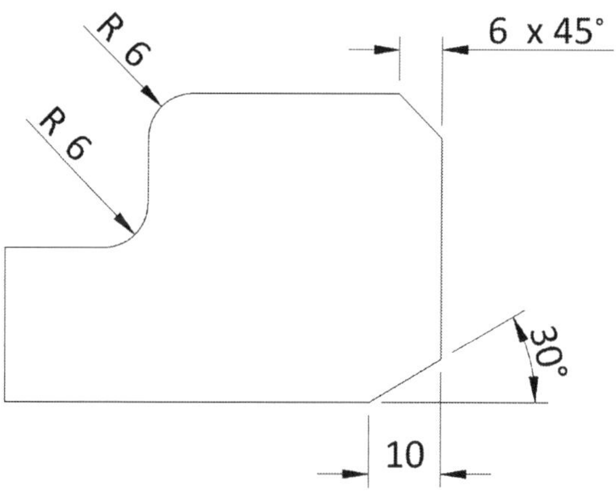

FIGURE 2.21 Chamfer and fillet dimensions.

It is also common to use fillets and chamfers as a general finish on all edges of components, and if so a popular method to show this is to use a callout or add a text comment on the drawing, for example; 'ALL EDGES FILLETED TO 5MM RADIUS' or 'ALL EDGES CHAMFERED TO 5MM × 45 DEG' or even 'ALL EDGES DEBURRED'.

2.11 CURVES AND SURFACES

Depicting **curve** properly can be a very challenging task. Like most drawing exercises the best method depends largely on the purpose of the output and the nature of the tolerances that must be adhered to. Good practice starts at the modelling stage. It's not necessary to understand all the maths behind curve representation to use curve commands in a CAD system but the more you do understand the better you'll be able to appreciate the commands and their limitations. Chapter 10 is dedicated to the representation of curves and surfaces in 3D modellers so you can use it or ignore it as you wish. For now we will consider only the common methods of dimensioning.

Perhaps the simplest method of representing curve geometry is to model with circular arc sections and straight lines. These can make for simple and clear drawings but have obvious drawbacks in terms of the geometry they can truly represent. The method is useful, however, in some application areas such as decorative feature design, or sometimes they are used in true representations of cam profiles. They are also often relatively convenient for CNC programmers to use since most NC controllers have linear and circular interpolation routines built in as G-codes (see Chapter 14). Figure 2.22 shows such a revolved surface section modelled with arcs and straight lines.

As an alternative, Figure 2.23 shows a curve with some coordinate point positions dimensioned. This form of dimensioning is no longer recommended in ISO

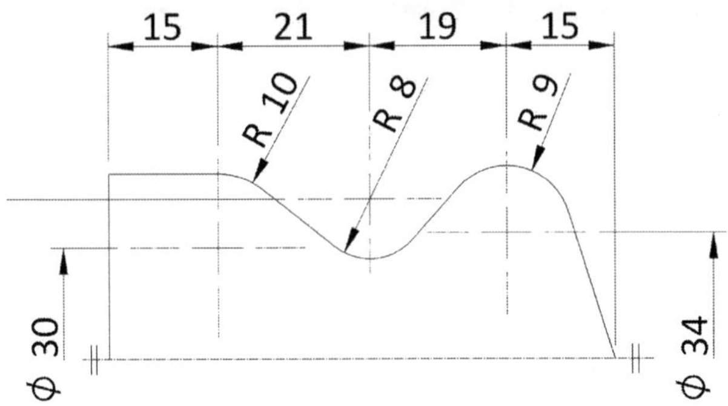

FIGURE 2.22 Revolved profile surface.

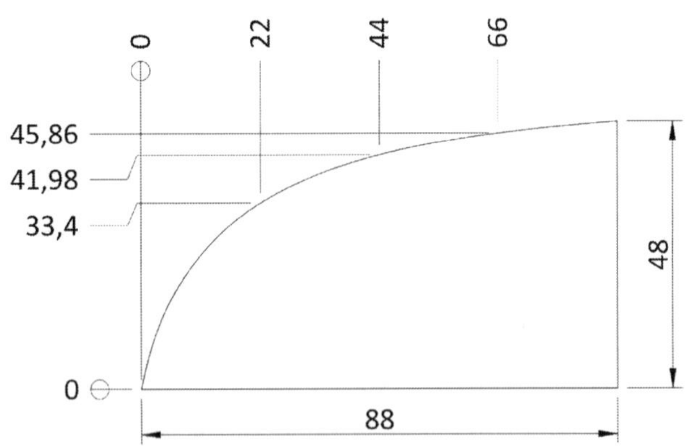

FIGURE 2.23 Spline curve by coordinate dimensions.

standards, but there is a great deal of legacy geometry that will be around for some time and people will continue to measure points from physical patterns or use their CAD systems to generate curves from an arbitrary selection of points that are to be interpolated. There are a number of shortcomings with such an approach.

1. It is common to generate curves in the original 3D model using arbitrarily placed points that are interpolated using specific algorithms and parameter conditions. These conditions may not be specified on the drawing, and the curve sections between the points are not fully defined.
2. There is a likelihood that some of the dimensions shown are rounded to 0, 1, 2 or more decimal places. If the curve is then recreated in a different application, the curve generated there might be significantly different from the original. Small differences in interpolation point positions can result in quite significant deviations in the curve geometry between the original and derived shapes.
3. There is a tendency to over-dimension curves in the belief that adding more coordinate points will lead to better or more accurate curves, but often the opposite is true. The fewest points used to fully define the desired shape should always be preferred. Adding unnecessary points can make downstream curve regeneration problematic.
4. CNC code generated from CAD curves is sometimes implemented using linear interpolation routines, and this means the curve might be cut on a machine with many very small linear segments. The geometry can be short of ideal and may be very slow to cut.

Increasingly, CAD packages and CNC controllers are offering better support for curve drawing through standard methods using common mathematical techniques based on Bézier, simple B-spline or NURBS-based modelling, for example.

When modelling surfaces, the problem of illustration on drawings is considerable more complex. Much more detail on the topic of curves and surfaces is given in Chapter 10.

2.12 THIN PARTS

Thin parts are often dimensioned in a slightly different way from common machined parts. For example, sheet metal boxes and housings are a regular feature of electronic and optical products and the processes used to make them are quite different from machined parts since they tend towards shearing and bending rather than cutting from solid stock. It is common on a drawing to depict the thickness of material using a special symbol. Figure 2.24 shows how this symbol is used.

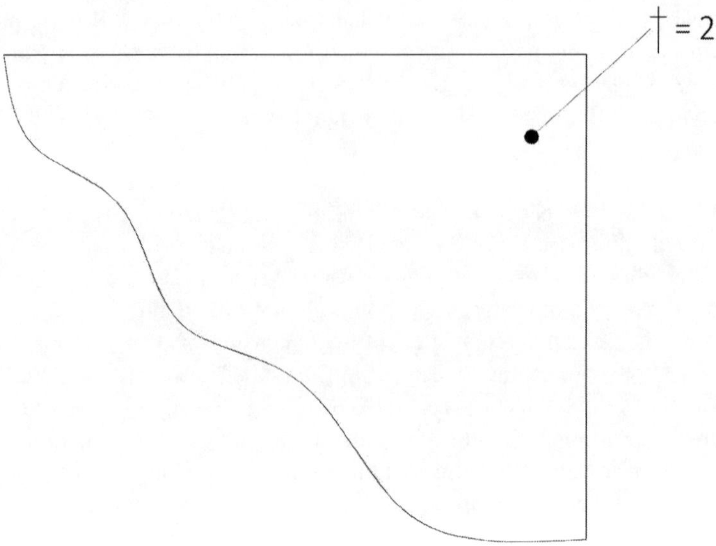

FIGURE 2.24 Dimensioning the thickness of a thin part.

2.13 DEVELOPED PARTS

Some parts are designed to change shape either temporarily (like springs) or permanently (like screw retainers), and it may therefore be necessary to depict the consequences of this on a drawing. Which condition (flexed or unflexed) should the part be shown in and what dimensions should be shown needs to be addressed. A part such as a spring might be shown in its free unconstrained state or in its constrained state. A simple example of a non-rigid part that is permanently deformed during manufacture can be depicted as a **developed part**, as shown in Figure 2.25, which shows a strip that is initially flat and then deformed by being bent around a cylindrical forming tool with a radius of 25 mm.

2.14 ISO STANDARD METHODS AND SYMBOLS

Before finishing a description of basic dimensioning techniques, it is worth pointing out that your ability to follow ISO standards may in part be constrained by what tools your CAD system supports for generating dimensions and callouts. It is common for CAD systems to implement a subset of the full set of ISO symbols, for example, or for the symbols given to be derived from earlier obsolete versions of a standard. Many companies augment their CAD systems with workarounds or with in-house written computer code to support their particular drawing practices.

Some of the common **property indicator symbols** used in dimensioning are shown in Table 2.1.

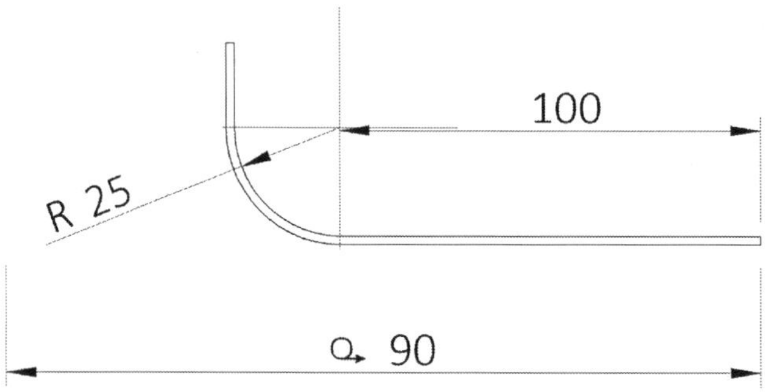

FIGURE 2.25 Developed part dimensioning.

TABLE 2.1
Special Symbols

Symbol	Meaning
⌵	Countersink
⌴	Counterbore
Ø	Diameter
↧	Depth
□	Square
R	Radius
SR	Spherical radius
SØ	Spherical diameter
⌒	Arc length
↔	Between
○➤	Developed length
t=	Thickness

2.15 COMMENT

Note: common symbols may be available from within the CAD system you are using, but it is worth checking these are up to date and accurate if you want to ensure compliance with a particular standard. Most of the symbols may be available in word processors using an add-in typeface such as IGES 1001, 1002 or 1003 or your CAD systems might install their own typefaces based on ISO symbols.

In the following chapter we will investigate initial methods used to add tolerances to dimensions, and then in the subsequent chapter we will consider more complex parts such as threaded, splined or serrated sections.

3 Introduction to Tolerancing

3.1 INTRODUCTION

The material covered in this chapter relates to methods for dimensioning sizes for real, that is non-ideal, shapes. A designer might have a perfectly straight-sided rectangular feature in mind but will know that the manufactured feature will not have a perfect shape or size and so must describe the acceptable limits of deviation to realize the design intent. The standards **ISO 14405–1:2016, ISO 17450–3:2016, ISO 286–1:2010** and **ISO 286–2:2010** describe ISO's recommendations on how we might go about dimensioning and **tolerancing** to achieve this purpose. Central to the concepts covered are the ISO terms, '**nominal**' feature as shown on a drawing and an '**integral**' feature as manufactured and measured. What follows in the first part of this chapter is an introduction to tolerancing in general, followed by an outline of some of the options on how these might be defined and measured and, subsequently, the method of using Limits & Fits is described.

A single dimension of length, even with a simple tolerance added, often implies a simplicity that does not stand up to even the most rudimentary scrutiny. Consider, for example, the drawing of a shaft in Figure 3.1.

What exactly do the dimensions mean? The length would appear to be 60 mm. Is that measured along the central axis? The outside edges? Do the dimensions imply anything about the shape? Can the cylinder actually be banana shaped? How round should the cylinder be? How perpendicular should the end faces be to the body? Or how parallel to each other? How 'smooth' do the surfaces have to be, and what is smoothness? What exactly is 'size'?

ISO 14405–1, in an attempt to address some of the issues with dimensioning, gives details of several methods of how a linear size might be defined. These can be grouped broadly as measures of **local size**, **calculated size**, **global size** or **rank order** size. Most engineers will be familiar with the most common method, that is '**two-point**' size, which is a local size, as illustrated in Figure 3.2.

The outline shown could be for an **internal** or **external feature of size** designed as an ideal cylinder—or if the ∅ symbol is not shown—a region between two parallel planes. Clearly there are an infinite number of pairs of points that could be used to measure across the profile, and therefore the designer may wish to indicate how the part should be measured to validate whether the feature will meet the specified tolerances. The example implies that several measures might be taken, and that each individual measure m1, m2, . . ., m7 would have to meet the tolerance shown. Two-point sizing is used in a variety of forms; sometimes, it is used across the maximum or minimum diameters of a cylinder, for example. However it is

DOI: 10.1201/9781003395072-3

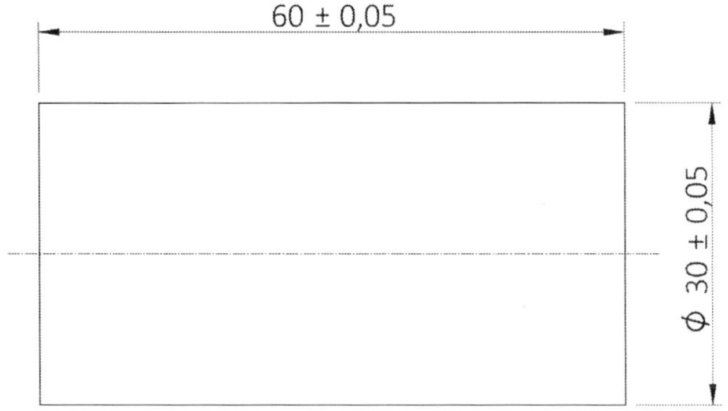

FIGURE 3.1 Simple tolerances on the shaft.

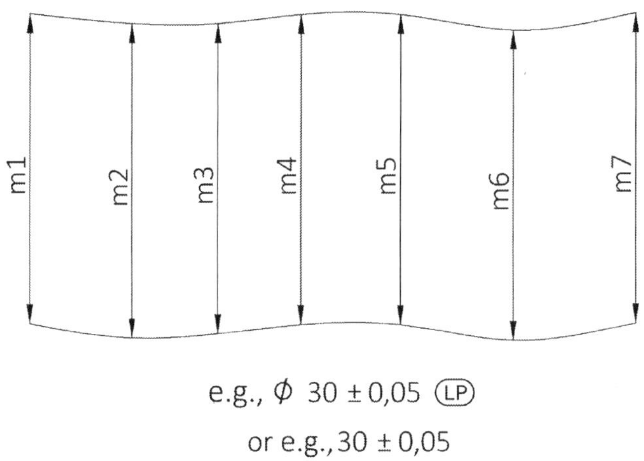

FIGURE 3.2 Two-point measures of size (local sizes).

used, it can be an ambiguous way to measure since it does not involve sampling the whole surface, and the size of the measurement 'points'—for example, the anvils on a micrometer—frequently impose a form of filtering effect on any profile on a surface. It is however the most common way for most engineers to measure and validate part dimensions and their associated tolerances on manufactured parts. This is the default measure of size in ISO 14405, so if the 'LP' modifier symbol (shown in Figure 3.2) is omitted, it can be assumed anyway.

The ISO standard also identifies another method of defining local size, as illustrated in Figure 3.3.

This method, the '**spherical size**', is a local size measure defined by the maximum size (diameter) of a sphere that will fit inside a cylindrical feature or one defined between two planes and at a given cross-section. The tolerance specification example shown in Figure 3.1 means that each spherical measure must be

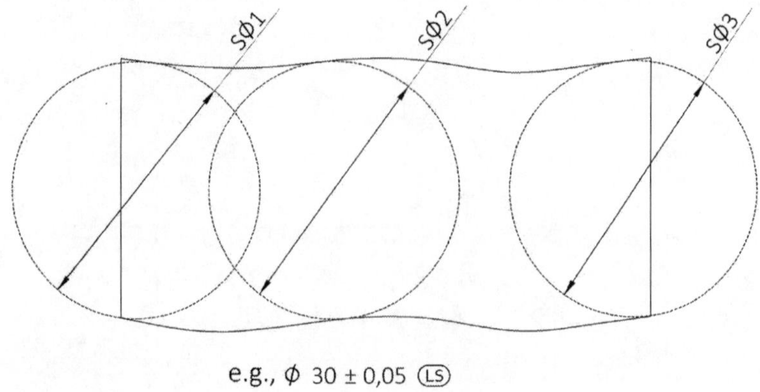

FIGURE 3.3 Local size defined by a sphere (spherical size).

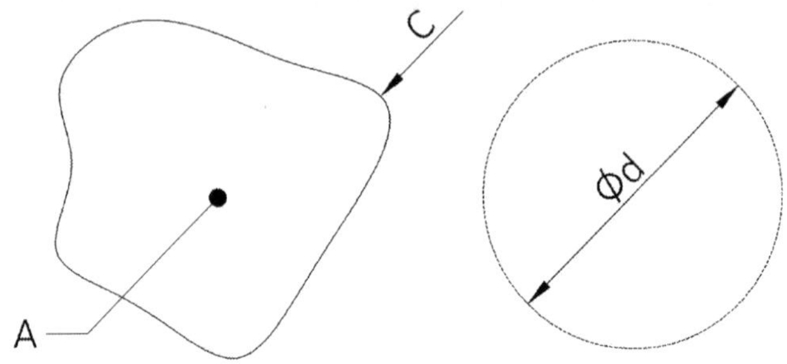

FIGURE 3.4 Cross-section.

within the tolerance bands stated, and this is invoked by the use of the '**LS**' modifier appended to the dimension/tolerance.

In addition to the local size direct measurements, it is possible to derive and calculate other dimensions from the direct measures taken. For example, if we consider a cylindrical feature in cross-section measured normal to the longitudinal axis, we might see a profile that looks like the one shown on the left in Figure 3.4.

There are several ways to calculate a diameter for such a feature in addition to simple measures of the outside diameter. Methods specified in ISO 14405 include:

Circumference diameter 'CC': $d = \frac{C}{\pi}$ where C is the length of the real feature profile

Area diameter 'CA': $d = \sqrt{\frac{4A}{\pi}}$ where A is the area of the cross-section

Volume diameter 'CV': $d = \sqrt{\frac{4V}{\pi L}}$, where versus is the average area × the length of the feature

Calculating circumference, area or volume can be difficult to do manually, but with modern metrology equipment such as laser scanners and touch probes with supporting software these measures are often readily available.

Also, more complex calculations of **global size** can be specified, and these are increasing in popularity. For example, Figure 3.5 illustrates graphically the concept of a feature of size being characterized by its maximum inscribed rectangular size.

This specification is invoked by the '**GX**' symbol.

Alternatively, a feature might be characterized by its **minimum circumscribed size**, as shown in Figure 3.6.

This specification is invoked by the '**GN**' symbol.

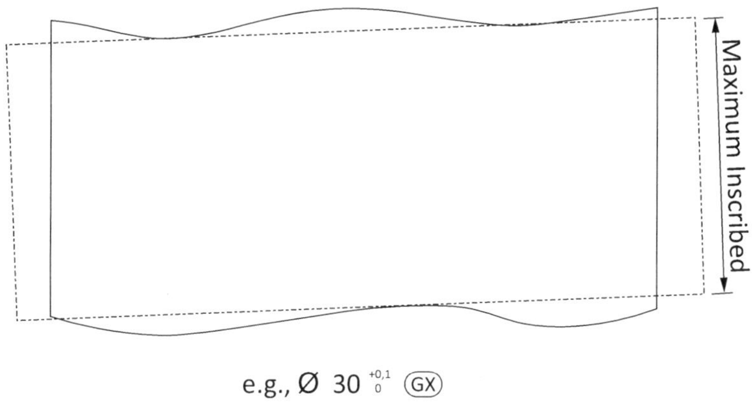

FIGURE 3.5 Maximum inscribed size (global measure).

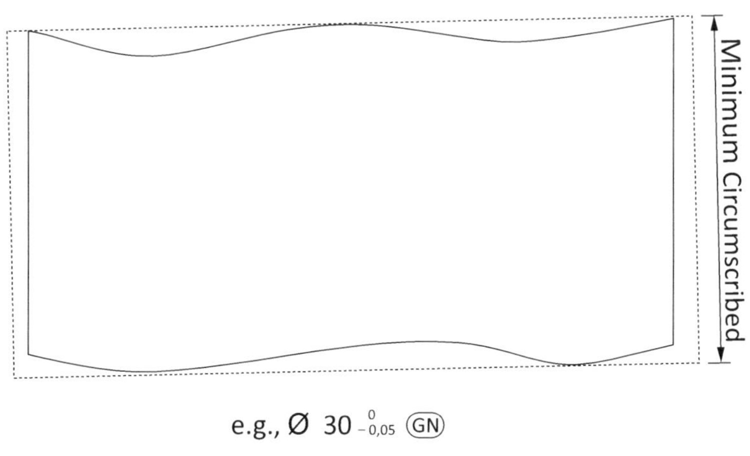

FIGURE 3.6 Minimum circumscribed size (global measure).

The **least-squares method** of measurement is illustrated graphically in Figure 3.7.

This measure, as defined in ISO 14405, involves fitting a rectangle horizontally in such a way that the top and bottom parallel lines minimize the square of the distance between them and the feature top and bottom profiles. It is invoked with the '**GG**' modifier symbol.

A more sophisticated way of fitting a rectangle to a feature of size involves allowing the rectangle to rotate to a given angle and minimizing the deviations between the rectangle and the profile top and bottom lines. This is known as the Chebyshev or minimax method and is illustrated in Figure 3.8 and is invoked by means of the '**GC**' modifier symbol.

Returning to the local size measures, it is often useful to use further calculated parameters that can be used to characterize aspects of the size of a feature, for example, the maximum size '**SX**' or the minimum size '**SN**' measured. Also, the average '**SA**', median '**SM**' and mid-range '**SD**' sizes can be readily found. It is also useful to specify sizes as a range of local sizes '**SR**' or to quote their allowable standard deviation '**SQ**'. Because the identification of median, mid-range and range parameters involves ordering a set of measurements according to their size, these measures are collectively referred to as **rank order** size methods.

A summary of the modifier symbols used to specify size is given in Figure 3.9.

These modifiers may be further detailed by adding additional complementary modifiers such as those relating to where measurements are applicable; for example, '**ACS**' would indicate that a tolerance dimension would apply at '**any cross-section**'. Further complementary specification modifiers are given in Figure 3.10.

Many of the measures outlined so far are most useful in association with geometrical tolerancing methods, which will be covered in later chapters; however, the common two-point method of measuring linear sizes is of particular utility in considering the use of limits and fits.

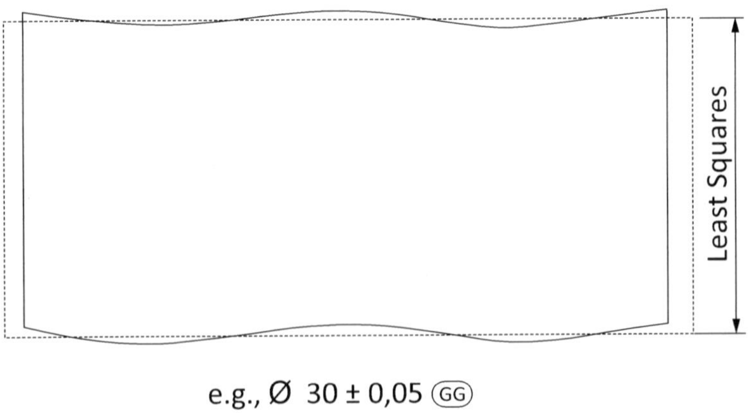

FIGURE 3.7 Least-squares size (global measure).

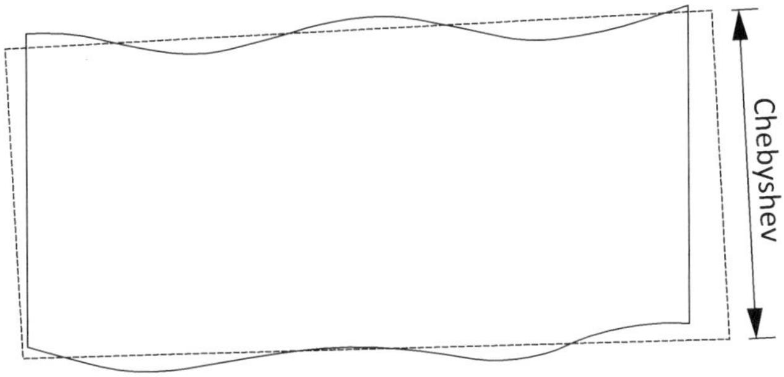

e.g., Ø 30 ± 0,05 (GC)

FIGURE 3.8 Minimax (Chebyshev) size (global measure).

(LP) Two point size
(LS) Local size defined by a sphere
(CC) Circumference diameter
(CA) Area diameter
(CV) Volume diameter
(GX) Maximum inscribed size
(GN) Minimum circumscribed size
(GG) Least squares size
(GC) Minimax size
(SX) Maximum size
(SN) Minimum size
(SA) Average size
(SM) Median size
(SD) Mid-range size
(SR) Range of sizes
(SQ) Standard deviation of sizes

FIGURE 3.9 Summary of modifiers for linear size.

UF United feature of size
Ⓔ Envelope requirement
ACS Any cross section
SCS Specific cross section
ALS Any longitudinal section
No. x More than one feature
CT Common toleranced feature of size
Ⓕ Free state condition
←→ Between
⊥ | B Intersection plane
⊥ | B Direction feature

FIGURE 3.10 Complementary specification modifiers.

3.2 MAXIMUM MATERIAL CONDITION AND LEAST MATERIAL CONDITION: MMC, MMS, MMR, LMC, LMS, LMR

When considering toleranced features, it is common to refer to the idea of a **maximum material condition (MMC)** and a **least material condition (LMC)**. This is illustrated graphically in Figure 3.11 for both an external feature (shaft) and an internal feature (hole).

The concept is simply that when an external feature such as a shaft or plate is at its upper tolerance limit, it will be at its largest size and therefore will be in its MMC. Conversely, when an internal feature such as a slot or hole is at its highest tolerance limit, it will be at its widest and therefore have the least material and mass. It is therefore in its LMC. Additionally, when it is at its lower tolerance limit, it will be at its narrowest or smallest size and therefore have maximum material and be at its MMC.

When an external feature of size is at its MMC, it will have the upper limit of its allowable size value, referred to in ISO standards as **maximum material size**, or **MMS**. Similarly, when an external feature of size is at its LMC, it will have the lower limit of its allowable size value, referred to in ISO standards as **least material size**, or **LMS.**

When a tolerance on a drawing is dependent on a feature being in its MMC, it is said to be required and is referred to as a **maximum material requirement**, or **MMR**. As you might guess, when a tolerance on a drawing is dependent on a feature being in its LMC, it is said to be required and is referred to as a **least material requirement**, or **LMR**. These can be important criteria in the specification of geometrical tolerances as we shall discover in future chapters.

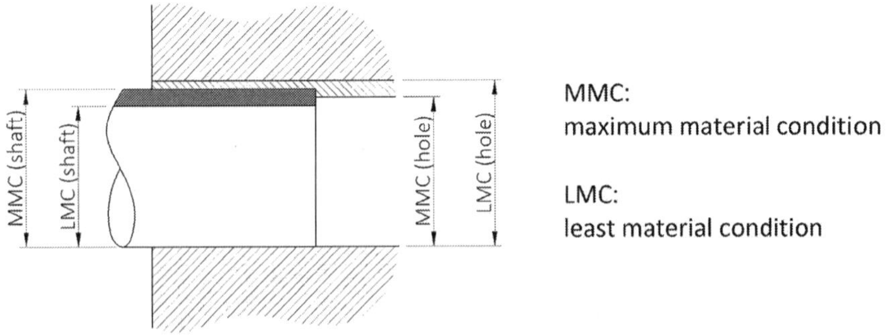

FIGURE 3.11 The maximum material condition and least material condition for toleranced hole and shaft.

3.3 ENVELOPES, DEPENDENCY AND THE TAYLOR PRINCIPLE

In the distant past, nearly all international standards (e.g. BS, DIN and ASME) interpreted simple sizes and their associated tolerances in the same way. This was based on what is widely known as the **Taylor principle**, the **envelope requirement** or the **dependency principle**. For example, an external feature such as a shaft of a nominal diameter of 20 mm might be depicted with a toleranced dimension of ∅20 mm ± 0,1. Any two-point measurements taken on the body's width (say m1–m7 in Figure 3.12) would have to be within the range of 19,9 mm – 20,1 mm. However, a set of parallel lines symmetrically distributed around a centreline (the envelope) would also be implied from this, and these would be 20,1 mm apart. The shaft, if within tolerance, would have a profile that could fully fit inside the lines.

When an external feature is at its MMC and has met the envelope requirement, it must have a perfect form; that is, it will fill the rectangular envelope completely. (In the ASME 14.5 standard this is often referred to as #Rule 1). If an external feature's tolerance refers to its MMC, then it gives rise to what is known as '**bonus tolerance**'. As the diameter or thickness of an external feature moves away from its MMC, more deviation from the perfect form can be allowed and the part will still fit inside its envelope (which is its minimum circumscribed size). The bonus realized will be the same magnitude as the deviation from MMC.

For an internal feature such as a slot or a hole, its envelope is the largest-sized part with perfectly parallel sides that will fit inside it, and at minimum tolerance or MMC, an internal feature must have the perfect form to meet the envelope

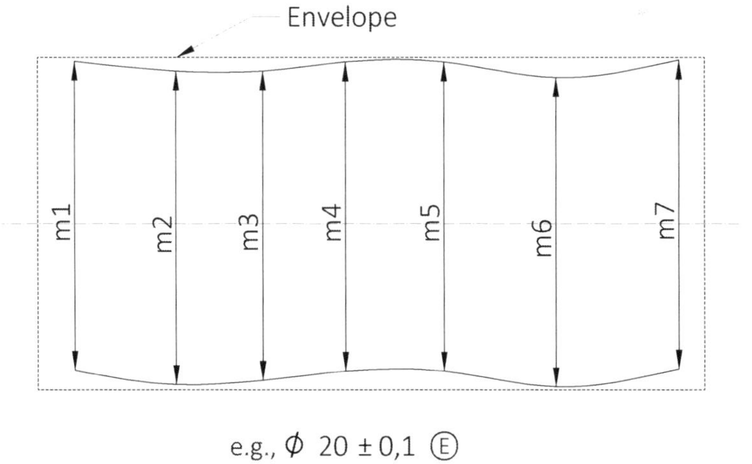

FIGURE 3.12 Dependant or envelope interpretation of a simple tolerance.

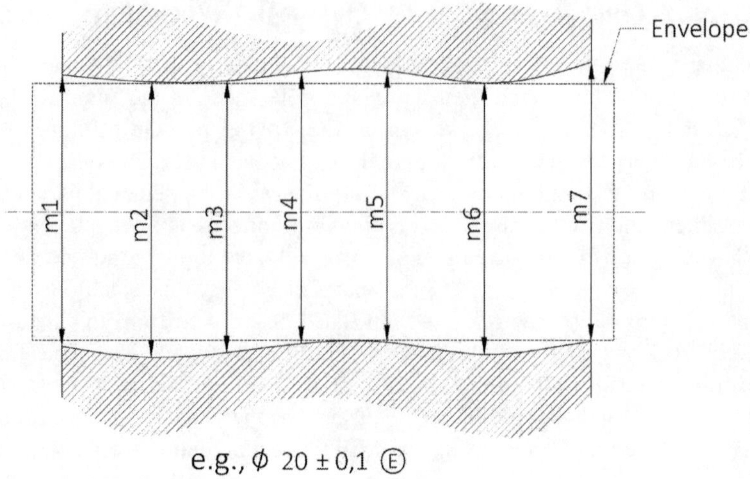

FIGURE 3.13 Internal feature envelope, for example, a hole.

requirement as shown in Figure 3.13. As the hole's actual diameter is widened to increased tolerance size and away from MMC, then increased deviation from the perfect form can be allowed and a bonus tolerance can be realized.

The diameters and gaps discussed are not always easy to measure and validate in production. Some form of specialist metrology equipment may be necessary and might require a complex set-up procedure. For diameters up to around 25 mm, there may be gauge pins that can be used for holes (pins that have extremely accurate form and typically come in sets with steps of 0,01 mm diameter between pins), and the largest pin that fits in a manufactured hole will represent the envelope size. For larger diameters it can become tricky, however, as vee blocks or other supports may need to accompany measurements with profile gauges. It is a similar situation for prismatic shapes.

One benefit of the envelope form of tolerance, especially when utilized in mass production, is that gauges (e.g. Go-NoGo) can be made to allow quick validation even at a workstation.

The envelope interpretation of a tolerance is often assumed, or not even thought about, by those new to engineering drawings, and it is important to consider an alternative way, using the **independency** principle as default in the ISO system.

3.4 INDEPENDENCE

A toleranced dimension such as 20 mm ± 0,1 does not carry any information about the form or shape of the part, and it can be interpreted as meaning that the tolerance can be validated simply by measuring the width of the shaft at opposing points along the profile as shown by the measurements m1–m7 in Figure 3.14. As long as all the measurements produce values between the tolerance limits 19,9–20,1 mm, the part

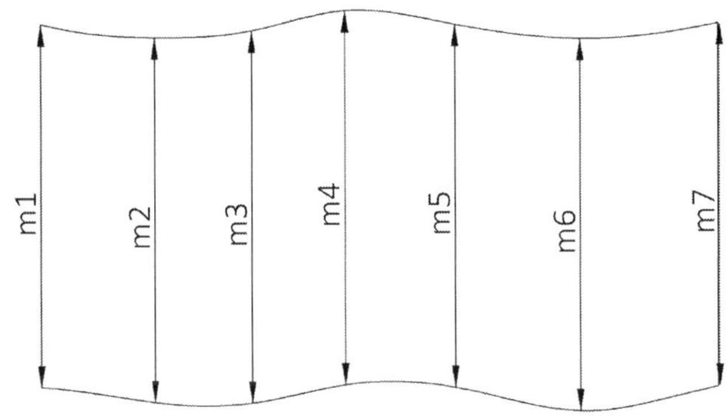

FIGURE 3.14 Independent interpretation of simple tolerance.

can be assumed to be acceptable. This might be as intended by the designer, but it is a potential fault on drawings to specify in this manner and end up with a part that may not be fit for its intended purpose; for example, the shaft may be curved or otherwise mis-shaped and cannot fit in a hole that is 20,1 mm wide.

As you can imagine the two different interpretations that can be implied from a simple tolerance dimension can give rise to serious misunderstandings and contractual wrangles. Clarity is therefore necessary, and when a drawing is declared to be adhering to a specific standard then the implications are clear, as follows:

In **ISO standards—independence** is assumed.
In **ASME 14.5-2018—dependency** is assumed.

Note that when a drawing invokes the ISO standards, it should be fully followed throughout the entire drawing and the whole geometry of the entire part should be specified with sizes and limitations of form and position (location and orientation). If the ISO standard is being used, it is possible to invoke dependency, or the envelope principle on an individual dimension by appending a specific symbol to it; the letter E is enclosed in a circle, as follows:

ϕ20mm +/– 0,1 Ⓔ

Thus, such a dimension means that all measurements along the part must be within 19,9–20,1 mm and that the part should fit inside perfectly parallel lines drawn 20,1 mm apart.

Alternatively, a general statement declaring dependency may be placed on the drawing near or in the title box stating:

LINEAR SIZE ISO 14405 Ⓔ

although it is also advisable to add a second statement clarifying what this statement means, for example:

ALL LINEAR SIZE DIMENSIONS SHOULD ASSUME THE ENVELOPE PRINCIPLE

3.5 COMMON DEPICTIONS OF TOLERANCES

Figure 3.15 shows the most common ways of expressing tolerances on individual length dimensions on engineering drawings. Most of these are self-explanatory; however, the **class/grade** method shown last, for example, **'30 g6'**, is explained in some more depth later in this chapter. Note that all of the tolerances shown should be interpreted with the principle of independence, as previously described.

Angular tolerances can be presented similarly with reference to nominal angular dimensions. While minutes and seconds can be used, it is common in many areas to use decimal measures, that is, angles in degrees and tenths, hundreds and thousandths of degrees, for example, 59,44°.

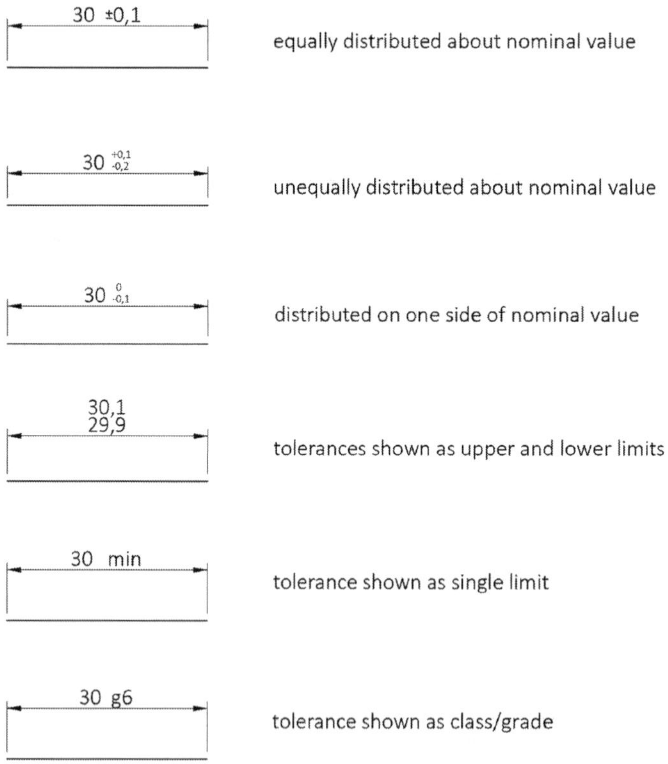

FIGURE 3.15 Common tolerances on lengths.

Tolerancing methods have probably been in use for a very long time, for example, as implied limits that were part of group or company policies; however, there is no known (by the author) reporting of explicit tolerances stated on drawings until around the early 1900s. Around 1902 a system developed at The Newall Engineering Company Limited for standardized sizing of matching shafts and holes was published. This was a shaft-based system whereby the diameters of standard shafts were used and different classes of tolerance were created for holes that were then created to give a required fit. This system, as a shaft-based system, was not popular with other industrialists because of the difficulty in making holes to fit specific dimensions. In most cases it is more convenient to use standard holes, based on common drill sizes, and then fit specific-sized shafts, since these can readily be turned on a lathe to any required dimension. A hole-based system was later developed for limits and fits, and a specification for these was released in 1924 by the British Standards Institution as BS164:1924.

3.6 LIMITS AND FITS WITH GRADES AND CLASSES

'**Limits and fits**' is a general term for a system of classifying suggested tolerance values for common parts that fit together around a common axis, for example, external surfaces such as solid cones, shafts and screws as well as their respective internal mating surfaces such as conical holes, cylindrical holes and female threads.

The full system of limits and fits is described in **ISO 286: 2010 parts 1, 2 and 3**. The tolerances described therein consist of a two-part designation: a letter describing the tolerance **class** and a number describing its **grade**, for example, 'g6' or 'H7'. The class describes the location of the tolerance, that is, how far it is from the nominal hole/shaft size, and the range describes how wide the tolerance value is. These ranges are recognized in the '**International Tolerance**' values and might be indicated as, for example, IT6, IT7 and so on, in tables of values. Hence 'g6' could represent a shaft with a 'g' class and tolerance range IT6. More of this later.

Note: Lowercase letters are used to indicate parameters relating to male, or external, surfaces while uppercase letters refer to most female, or internal, surfaces.

The ISO 286 standard is very wide-ranging in its coverage and can be quite demanding to use, especially if you do not use limits and fits frequently. Throughout the following discussion, details will be kept to a minimum and selected examples will be used to show how the system works without burdening the reader with detail that can be referred to if needed, either through web-based resources or by referring to the full ISO standards.

Using shafts and hole assemblies as our primary example, we can first consider what is meant by fits before exploring the limits that might be imposed by tolerance values. The ISO standards refer to three very broad categories of fit: **clearance, transition and interference**. Figure 3.16 shows a graphical representation of the three fits as described in ISO 286.

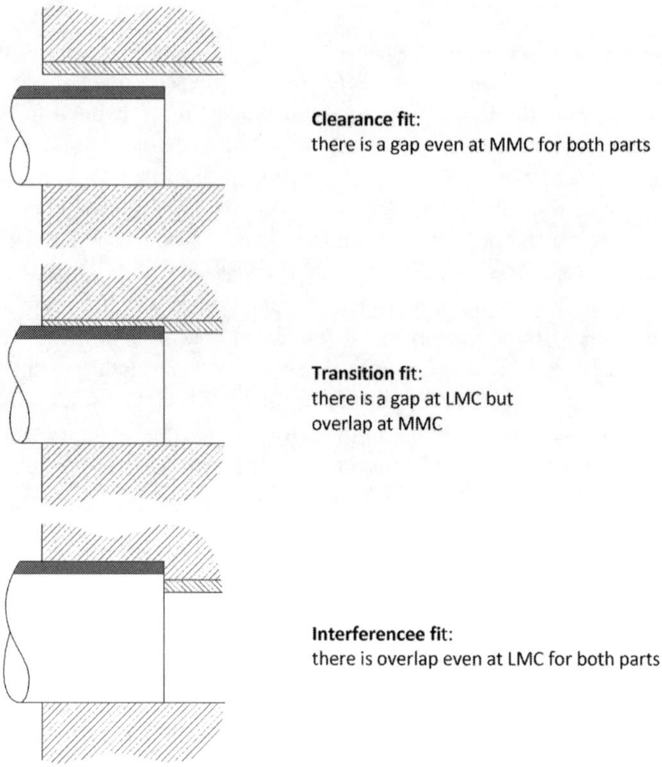

FIGURE 3.16 Types of fit.

Clearance fits are those fits where a shaft and a hole will always have a theoretical gap between them when assembled (although sometimes the gap = 0 mm for some designations, as we shall see). These fits are useful and commonly employed for simple bearing surfaces and coarse positioning applications. They vary considerably in terms of how large a gap might be, and they can be further broken down into subcategories such as '**Easy Running**', '**Close Running**' and '**Sliding**'.

Transition fits allow some overlap in size between the shaft and the hole; for example, when the hole tolerance has the hole diameter at its smallest value (MMC) and the shaft is at its largest possible diameter (MMC), the parts will not likely fit together without an applied force, due to their overlapping dimensions. When the hole is at its largest diameter (LMC) and the shaft at its smallest diameter (LMC), there will be some clearance, and assembly will require no substantial force. This type of fit is used in applications where locational accuracy is desired (more so than with a clearance fit), but excessive stresses are to be avoided as these may impair performance unduly. These fits are sometimes further broken down into subcategories such as '**Push Fit**' or '**Drive Fit**'.

Interference fits are used where an overlap of the basic sizes is desired, and there will be some overlap at any allowable value of the tolerances, even when both parts are at LMC. These types of assemblies commonly require special equipment to assemble the parts and may include powered pressing facilities and/or relative heating of the internal surfaces to ensure suitable expansion to allow assembly. A very common historical application for interference fitting was the assembly of steel tyres on the wheels of steam locomotives. Interference fits are also useful for inserting bearing bushes. These fits are sometimes further broken down into subcategories such as '**Light Press Fit**' or '**Press Fit**'.

3.7 USING ISO 286-1:2010 AND ISO 286-2:2010

Although there are many different possible use cases for applying limits and fits there are also a limited number of ways in which most users will adopt the standard. Before accessing the various tables and information needed to specify limits, the user might want to decide what sort of fit might be relevant and whether to start from a **'hole-based'** approach to specifying the tolerance limits on a shaft or whether to start with a **'shaft-based'** approach and then specify tolerance limits on the hole dimensions.

3.7.1 Hole-Based versus Shaft-Based

As has been stated, in most cases, at the outset, a designer will decide whether to use a 'hole-based' or 'shaft-based' system to work on. The ISO 286 standard recommends using hole-based methods for most cases, mainly because it is usually relatively easy to create a standard-sized hole by drilling/reaming and then turning/grinding a shaft to the given tolerance to ensure the desired fit. For simplicity we shall assume a hole-based approach for now; however, the use of a shaft-based system will follow. And for further clarity we will look at a subset of the class/grade types given in the ISO standards, but it will be sufficient to show how the system works and is useful for the most common use cases.

Figure 3.17 shows graphically, a small selection of tolerance grades for holes and how they relate to the nominal size of a feature.

where (from ISO286:1):

EI = lower tolerance value
ES = upper tolerance value
IT = tolerance grade range
Δ = adjustment, only used for classes K, M and N with diameters up to 500 mm and grades up to IT8, and for classes P through to ZC with grades IT3-IT7,

ES = EI + IT (values for either EI or ES as well as IT are obtained from tables)

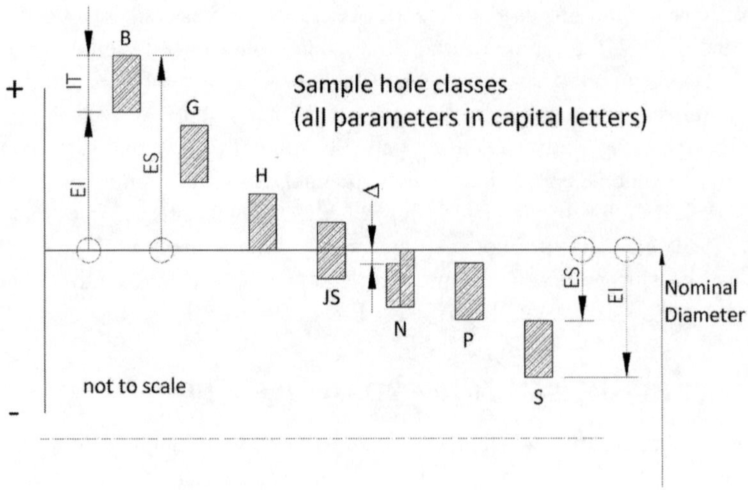

FIGURE 3.17 Sample hole classes.

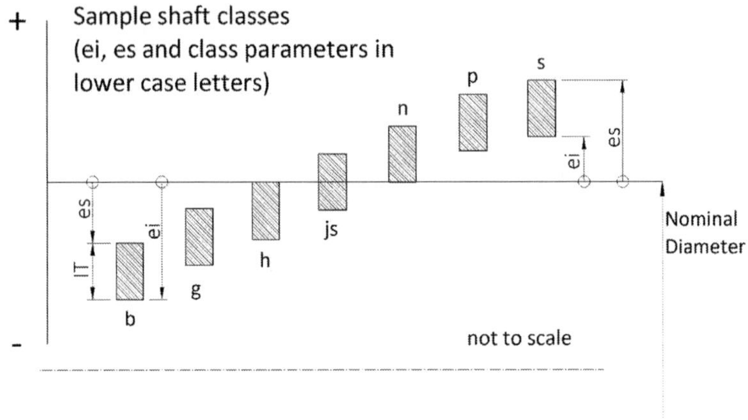

FIGURE 3.18 Sample shaft classes.

Similarly, sample shaft classes can be shown graphically as in Figure 3.18.
where (from ISO286:1):

ei = lower tolerance value
es = upper tolerance value
IT = tolerance grade range
es = ei + IT (values for either ei or es as well as IT are obtained from tables)

The ISO 286 series of standards describes a very comprehensive set of possible classes and grades of tolerancing for a wide set of applications. For example, the alphabetic designation of classes consists of 28 different letters or letter combinations, while there are some 20 different ranges (IT01, IT0, IT1, . . ., IT18). IT01 represents the finest tolerance range, and IT18 the widest or most coarse. The standard does, however, recognize that for most users only a subset of the data provided might be used and a set of preferred hole/shaft classes and grades are as follows:

Preferred hole-based pairings

H7: g6, h6, js6, k6, n6, p6, r6, s6 (e.g. H7/g6, H7/h6 . . . . . H7/s6)
H8: e8, f7, h7
H9: e8
H11: b11, c11

Preferred shaft-based pairings

h6: G7, H7, JS7, K7, N7, P7, R7, S7
h7: F8, H8
h9: B11, D10, E9, F8, H8, H9

For the most part, it is unlikely that a designer would want to specify pairings that are both largely positive or largely negative and classes are usually grouped around the nominal sizes, although sometimes it is useful to use pairings where both hole and shaft are both positive or both negative.

It is also of value to pair tolerances with similar ranges; for example, a very coarsely defined hole would not often be paired with a very finely defined shaft so the range values used are usually reasonably close to each other.

In fact ISO 286 recommends that only 12 hole classes need to be used 'for everyday engineering' purposes and only 12 shaft classes. We will concentrate further discussion on these for nominal diameters up to 180 mm. What follows are three tables of data from which all tolerancing limits can be calculated for all of the preferred depictions stated in ISO286 for sizes up to 180 mm.

Table 3.1 shows values of EI and ES for selected hole classes and diameters up to 180 mm.

Table 3.2 shows values of ei and es for selected shaft classes with diameters up to 180 mm

Table 3.3 shows a small selection of values for commonly used grades IT6-IT11.

To find the numerical tolerance dimension values we use the formulae.

$$ES = EI + IT \text{ and } es = ei + IT$$

TABLE 3.1
Selected Hole EI or ES Values

Nominal Size (mm)		EI (μm) for Grades Shown							ES (μm) for Grades Shown					Values for Δ for Selected Grades (μm)			
Above	Up to	B All Grades	D All Grades	E All Grades	F All Grades	G All Grades	H All Grades	JS All Grades	K Up to IT8	N Up to IT8	P Up to IT7	R Up to IT7	S Up to IT7	IT5	IT6	IT7	IT8
0	**3**	140	20	14	8	2			0	−4	−6+Δ	−10+Δ	−14+Δ				
3	**6**	140	30	20	10	4			−1+Δ	−8+Δ	−12+Δ	−15+Δ	−19+Δ	1	3	4	6
6	**10**	150	40	25	13	5			−1+Δ	−10+Δ	−15+Δ	−19+Δ	−23+Δ	2	3	6	7
10	**14**	150	50	32	16	6			−1+Δ	−12+Δ	−18+Δ	−23+Δ	−28+Δ	3	3	7	9
14	**18**	150	50	32	16	6			−1+Δ	−12+Δ	−18+Δ	−23+Δ	−28+Δ	3	3	7	9
18	**24**	160	55	40	20	7			−2+Δ	−15+Δ	−22+Δ	−28+Δ	−35+Δ	3	4	8	12
24	**30**	160	55	40	20	7			−2+Δ	−15+Δ	−22+Δ	−28+Δ	−35+Δ	3	4	8	12
30	**40**	170	80	50	25	9	0	–IT/2	−2+Δ	−17+Δ	−26+Δ	−34+Δ	−43+Δ	4	5	9	14
40	**50**	180	80	50	25	9			−2+Δ	−17+Δ	−26+Δ	−34+Δ	−43+Δ	4	5	9	14
50	**65**	190	100	60	30	10			−2+Δ	−20+Δ	−32+Δ	−41+Δ	−53+Δ	5	6	11	16
65	**80**	200	100	60	30	10			−2+Δ	−20+Δ	−32+Δ	−43+Δ	−59+Δ	5	6	11	16
80	**100**	220	120	72	36	12			−3+Δ	−23+Δ	−37+Δ	−51+Δ	−71+Δ	5	7	13	19
100	**120**	240	120	72	36	12			−3+Δ	−23+Δ	−37+Δ	−54+Δ	−79+Δ	5	7	13	19
120	**140**	260	145	85	43	14			−3+Δ	−27+Δ	−43+Δ	−63+Δ	−92+Δ	6	7	15	23
140	**160**	280	145	85	43	14			−3+Δ	−27+Δ	−43+Δ	−65+Δ	−100+Δ	6	7	15	23
160	**180**	310	145	85	43	14			−3+Δ	−27+Δ	−43+Δ	−88+Δ	−108+Δ	6	7	15	23

TABLE 3.2
Values of ES and EI for Selected Shaft Classes

Nominal Size (mm)		es (μm) Valid for IT Ranges Shown							ei (μm)				
Above	Up To	b for IT 8–12	c for IT 8–12	e for IT 5–10	f for IT 4–9	g for IT 4–8	h for IT 1–18	js for IT 1–18	k for IT 4–6	n for IT 4–7	p for IT 4–8	r for IT 4–8	s for IT 4–9
0	**3**	−140	−60	−14	−6	−2			0	4	6	10	14
3	**6**	−140	−70	−20	−10	−4			1	8	12	15	19
6	**10**	−150	−80	−25	−13	−5			1	10	24	19	23
10	**14**	−150	−95	−32	−16	−6			1	12	29	23	28
14	**18**	−150	−95	−32	−16	−6			1	12	29	23	28
18	**24**	−160	−110	−40	−20	−7			2	15	35	28	35
24	**30**	−160	−110	−40	−20	−7			2	15	35	28	35
30	**40**	−170	−120	−50	−25	−9			2	17	42	34	43
40	**50**	−180	−130	−50	−25	−9	0	IT/2	2	17	42	34	43
50	**65**	−190	−140	−60	−30	−10			2	20	51	41	53
65	**80**	−200	−150	−60	−30	−10			2	20	51	43	59
80	**100**	−220	−170	−72	−36	−12			3	45	59	51	71
100	**120**	−240	−180	−72	−36	−12			3	45	59	54	79
120	**140**	−260	−200	−85	−43	−14			3	52	68	63	92
140	**160**	−280	−210	−85	−43	−14			3	52	68	65	100
160	**180**	−310	−230	−85	−43	−14			3	52	68	68	108

TABLE 3.3
Selected IT Range Values

Nominal Size (mm)		Standard Tolerance Values from Selected Grades (μm)					
Above	**Up To**	**IT6**	**IT7**	**IT8**	**IT9**	**IT10**	**IT11**
0	**3**	6	10	14	25	40	50
3	**6**	8	12	18	30	48	75
6	**10**	9	15	22	36	58	90
10	**18**	11	18	27	43	70	110
18	**30**	13	21	33	52	84	130
30	**50**	15	25	39	62	100	160
50	**80**	19	30	46	74	120	190
80	**120**	22	35	54	87	140	220
120	**180**	25	40	63	100	160	250

3.8 EXAMPLES

EXAMPLE 1

A hole dimension might be depicted as follows:

ϕ 20 G7Ⓔ
from Table 3.1 in column G row 18–24, EI = 7 μm
from Table 3.3 in column IT7 row 18–30, IT = 21 μm
therefore,
the upper tolerance ES = 7 + 21 = 28 μm
the lower tolerance EI = 7 μm.
giving $\varnothing \mathbf{20}^{+0{,}028}_{+0{,}007}$ Ⓔ as the equivalent tolerance in numerical format.

EXAMPLE 2

A designer chooses to specify a hole/shaft assembly with a hole-based transition fit using an ISO-preferred pairing as follows:

ϕ 10 H7Ⓔ/js6Ⓔ
For the hole we find
from Table 3.1 column H, EI = 0 (in fact *all H sizes have EI = 0*, so we don't really need the table)
from Table 3.3 column IT7 rows 6–10, IT = 15 μm
giving $\varnothing \mathbf{10}^{+0{,}015}_{0}$ Ⓔ as a numerical tolerance for the hole, and
from Table 3.2 column H (any row) es = IT/2
from Table 3.3 rows 6–10, IT6 = 9 μm, es = 4.5 μm
giving $\varnothing \mathbf{10}^{+0{,}0045}_{-0{,}0045}$ Ⓔ as a numerical tolerance for the shaft.

EXAMPLE 3

Using a shaft-based preferred ISO pairing the following has been specified:

ϕ 30 N7Ⓔ/h6Ⓔ

from Table 3.1 column N, rows 24–30, ES = −15 + Δ μm and from column IT7, Δ = 8 μm

therefore ES = −15 + 8 = −7 μm

from Table 3.3 column IT7 rows 18–30, IT = 21 μm

and EI = ES–IT so EI = −7–21 = −28 μm

giving $\varnothing\mathbf{30}^{\mathbf{-0,007}}_{\mathbf{-0,028}}$ Ⓔ as an equivalent numerical tolerance for the hole.

from Table 3.2 column h, es = 0 (in fact *all h sizes have es = 0*, so we don't really need the table)

from Table 3.3 column IT6 rows 18–30, IT = 13 μm

giving $\varnothing\mathbf{30}^{\mathbf{0}}_{\mathbf{-0.013}}$ Ⓔ as an equivalent numerical tolerance for the shaft.

For clarity it may also be convenient to show the tolerance pairings on assembly drawings as shown in Figure 3.19.

The examples given show how the system of limits and fits uses preferred pairings of tolerance classes and grades for common internal and external features, and it is clear that this system of specification can be equivalently depicted using standard numerical tolerances. The use of preferred standard pairings, however, may be found to be both convenient and hence cheap to use. For example, standard engineering components such as nuts and bolts are frequently advertised with limits and fits for the associated thread parameters. See also Chapter 4.

At its simplest, the use of limits and fits might allow a designer to choose a nominal hole/shaft size and then decide on what sort of fit might be appropriate; for example, 'H7/g6' would specify a close running fit. The manufacturer can then convert these specifications into numerical tolerance values. Alternatively, some suppliers of standard parts will supply parts with the alphanumeric designations of their components shown in their technical data sheets.

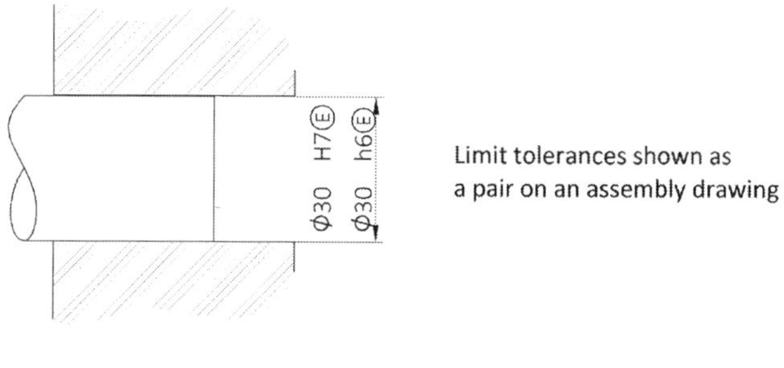

FIGURE 3.19 Limits on an assembly drawing.

3.9 COMMENTS

The dichotomy between the use of the principles of dependence and independence left many questions open. Even with dependency, for example, there is still considerable ambiguity on what is meant by size. Toleranced dimensions may seem even more ambiguous; however, the assumption with their use is that any indications of size will be augmented with full constraining information on the feature's form, location and orientation. The method of doing this is widely called geometric tolerancing, and the standards organizations refer to the full scope of their specifications in the following way. ASME through the ASME 14.5 standard refers to their system as **geometric dimensioning and tolerancing, GD&T**, while the ISO set of standards (e.g. ISO 1101) refers to their system as **Geometric Product Specification, or GPS**. The two sets of standards have some differences in the underlying assumptions, or axioms, that they use, but they broadly use similar methods to constrain part specifications on engineering drawings. The GPS standard places a little more emphasis on the validation aspects and results in differences in how sizes might be calculated. It also includes specifications for surface texture and edge tolerances.

4 Threads, Splines and Knurling

4.1 INTRODUCTION

There are many situations where basic views and dimensioning methods are not appropriate for specific feature types. Some features have complex geometry, and this can affect how parts are modelled, how engineering drawing views are constructed and how appropriate dimensions are shown. Best practice, as always, is largely dependent on the purpose of the output, whether that be part of the documentation in the form of an enhanced 3D model, an engineering drawing or other **product manufacturing information (PMI)**-based output.

Perhaps the most common scenario to consider is how the part is planned to be produced. For example, if a specific threaded part is to be manufactured, possibly in-house or sub-contracted, it may be necessary to provide complete geometric details of the finished part and any **intermediate forms**. This may include full 3D modelling and drawings complete with full geometry and allowable variation (tolerance). These might also be needed if **photo-realistic renderings** of a part are required. However, if the thread is a standard shape it may be preferable to provide simpler drawings depicting the necessary standard parameters for the thread geometry together with additional technical data that may be required for tool selection or machine set-up. In both cases sufficient detail should be given so that the final parts can be validated by reference to the requirements shown on the drawing or in the agreed-upon technical data specified.

In the case of **bought-in standard parts** it may be sufficient to specify the component type and supplier and then the design and validation might be based on supplier-produced data sheets specifying shape, materials, dimensions and tolerances. Many companies use some form of statistical-based **acceptance sampling** for such parts.

Let's consider some example situations.

In most cases threads are used in standard configurations. These can be cut in **lathe-type machines**, often with tools specifically made for efficient processing, or they may be made by forming-type operations, cut with **taps and dies** or by 3D printing. **Threaded inserts** are also very commonly used on parts made from softer materials such as wood or plastics.

It is useful to understand some of the basic means by which common thread types are designed and specified. Firstly, screw threads can be broken down into four main and commonly used types, **ordinary screw**, **buttress**, **acme** and **square threads**. Most common fastener types make use of a standard ordinary screw thread, while power transmission applications often use buttress, acme or square threads. All

DOI: 10.1201/9781003395072-4

threads have mating helical surfaces. The external (or male) thread, for example, a bolt, has its thread on the outside (or external) part of its body, while the internal (or female) thread, for example, a nut, has the thread part on the inside of its body.

On an engineering drawing or PMI depiction, it may be necessary to use one or more of a variety of means of showing the design intent. It's desirable to choose the clearest depiction that fully encompasses the needs of the design and that is usually the simplest one that suits this purpose. It's important to understand the meaning and purpose of the terminology used in thread descriptions and we can begin by considering the most common metric thread type: the ISO 'M' series, for example, M6, a standard 6 mm nominal diameter thread.

HISTORICAL NOTE

Helical configurations have formed the basis of artefacts for a very long time, for decoration (jewellery), doing work (water screws) or simply for fun (helter-skelter). In fact helical structures are as old as life itself (e.g. DNA) partly because they can be configured to neatly fold shapes into efficient packing regimes (e.g. in folding flexible tubes used in food packaging applications such as sausage skins).

The helices we will discuss in this book, however, will be those relating to engineering screws and particularly threads of various types. Nearly all such thread arrangements have in common the fact there is one turning part and one non-turning part at any time. Applying the appropriate torque to one part will induce a linear motion in one of the two parts. Thus, screw threads are used to convert torque to linear force. Screw principles were known to be used around 500 bce and were depicted by Archimedes around 400 bce. They were widely drawn with several applications by Leonardo Da Vinci. Typical early applications were commonly used for water transport in bilge pumps on ships, in irrigation and in grape presses.

As far as we know early thread regimes had little or no standardization, so individual manufacturers could make their own thread patterns, and usually did. Threads were known to be used in Germany in the 1500s, and early screw-cutting machinery was evident from around the mid-1500s. The earliest known and widely documented standard thread configuration in common use was promoted by Joseph Whitworth in England during the Industrial Revolution. In 1841 he proposed a common thread angle of 55° and a series of preferred threads per inch for the various diameters. Together with useful screw-cutting lathes which had been developed in England in the early 1800s. This made it possible to make standard threaded components which could be assembled with matching threads and this gave rise to nuts and bolts.

Thread types have been widely developed since then, and we now have a plethora of standards, which is convenient or inconvenient depending on your point of view.

4.2 SCREW THREADS

The most common type of thread in everyday use is the screw thread, as is used in nuts and bolts and a host of other applications. There is a plethora of different thread types available to the engineer, and some knowledge of the varieties is useful because there are several standards in current use around the world, and historic thread forms must also be catered for on occasion. For the 'metric' engineer, for example, knowledge of historic imperial threads is commonly needed if the design is for the maintenance of older equipment or to design new equipment that needs to connect to imperial-based threads. Perhaps the most common everyday example is metric-based equipment that is made to support cameras. These normally have a mounting thread that is expressed as an **ANSI UNC** type ¼″ × 20 tpi (threads per inch). Heavier professional equipment is usually fitted with a 3/8″ × 16 tpi mounting hole. These can be designed/specified in metric form according to an **ISO 1222:2010** standard but these have diameter sizes of 6,35 and 7,9375 mm, respectively.

What follows is a description of some of the most common thread types, and the first of these, the **ISO 'M' series**, will be covered in some depth. Many of the points covered can then be related to the coverage of other thread types using such terms as **ANSI, Unified, Whitworth, BS** and **BA**, for example.

4.3 'M' SERIES

Figure 4.1 shows an idealized (and exaggerated) typical thread form for an ISO-type standard metric or 'M' thread based around an external (male) threaded shaft (e.g. a bolt).

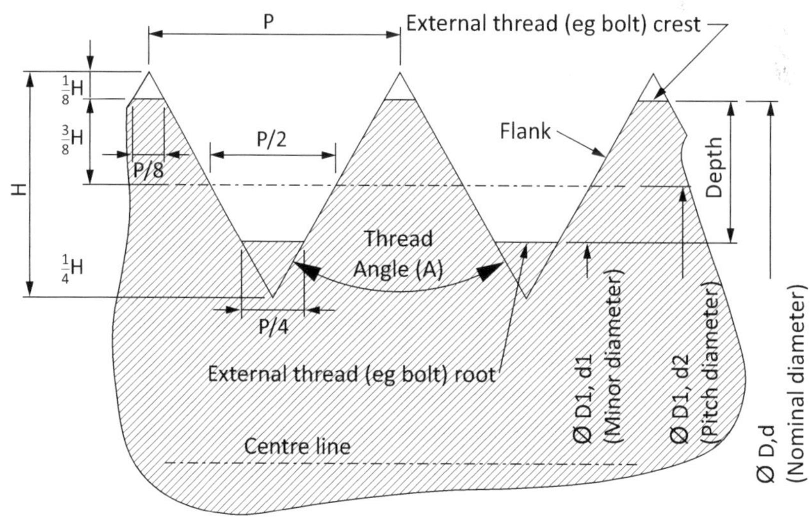

FIGURE 4.1 M-type screw thread.

Note that the thread is formed around the construction geometry of a simple sawtooth shape and has various parameters that are used to define the nominal thread (shown in bold), for example:

Pitch (P)—the distance between successive thread forms

Height (H)—this is the overall height of the construction shape used to create the thread. From simple trigonometry;

$$H = \frac{Tan\,A}{2}P$$

and for the ISO 'M' system, since A = 60°;

$$H = \frac{\sqrt{3}}{2}P$$

and from this equation and our selection of nominal thread diameter we can derive all of the other variables shown.

Crest—the flattened area at the outermost edge of the external (or male) thread, for example, a bolt. Most threads have such a flattened or rounded area to enable suitable mating of thread forms, aid manufacture and simply avoid unwanted potentially hazardous sharp edges. For an internal thread such as a nut the crest is the part of the thread at the minor diameter.

Root—the flattened area at the innermost part of a thread. In practice this is also rounded in some way.

Depth—this is the distance between the crest and root of the thread.

D, d—These are based on the nominal size of the thread, known as the major diameter. *D* is the basic major diameter of the internal thread (e.g. the nut) while *d* is the major diameter of the external thread (e.g. the bolt). For the idealized non-toleranced thread form shown, $D = d$.

D1, d1—These are the basic minor diameters of the internal and external threads, respectively, and these too are based on the nominal profile. They are derived from the minor diameter of the thread.

D2, d2—These are the pitch diameters of the internal and external threads, respectively, and these too are based on the nominal profile. They represent the width of the thread at the pitch diameter.

Flank—the sloping or load-bearing section of the thread.

Thread Angle (*A*)—the angle between successive flanks. The flanks are usually symmetric for a screw thread, and in the ISO system they sit at 30° to the vertical. This gives a standard thread angle A = 60° in the 'M' system and several other standards.

Values for the various variables shown can be accessed from the relevant standards, for example, from general screw thread information in **ISO 68, ISO 261, ISO 724** and associated tolerance data in **ISO 965**. From simple trigonometry, however, we can devise all of the variable values from the pitch and the nominal thread diameter.

In most applications, such as ordinary nuts and bolts, the type of screw thread most commonly used in the metric ISO system is a **coarse thread**. **Fine threads** have smaller pitch values than coarse. Note that the terms 'coarse' and 'fine' refer only to pitch values and do not represent any implied manufacturing or finishing attributes of a thread in this context.

Using the ISO coarse thread system, a nominal diameter can be chosen for a specific application, say 10 mm, and from the ISO standards mentioned, an M10 thread type is specified with a pitch of 1,5 mm. From these values, all of the other attributes such as crest, root, depth and pitch diameter are derived.

The need to ensure reasonable manufacturing methods, ease of assembly, functional integrity and allowance for any surface treatments and coatings means that threads cannot normally be made according to the nominal or fundamental geometry, and we must consider the use of deviations from the nominal as well as tolerances and the introduction of some radii to give actual complete dimensions of practical thread profiles. Figure 4.2 shows a **nominal thread profile** together with small gaps between the nominal profile and the profile of the internal thread and between the nominal profile and the outside form of the external thread.

The gap between the internal and external thread represents a clearance that must be allowed to obtain the required fit. Note also that internal sharp corners are often rounded, and these form radii such as the **root radius** for the bolt. If an exact value of radius is to be specified, it is common that the arc will be tangential to the thread flanks, and as before this value can be calculated from the pitch for a recommended value, as follows.

$$\text{Bolt Root Radius} = \frac{\sqrt{3}}{12}P$$

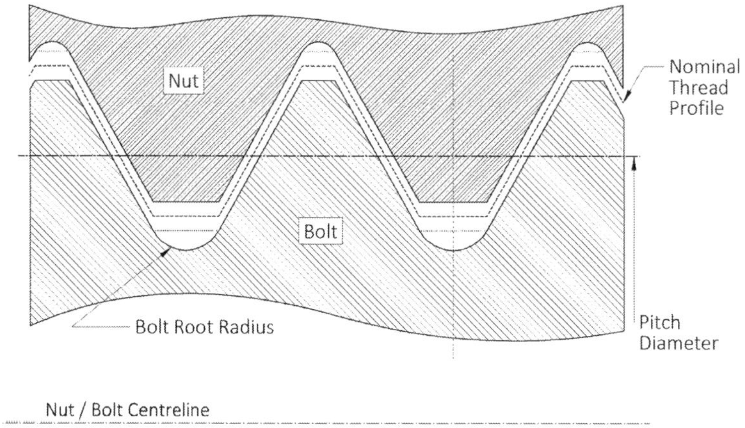

FIGURE 4.2 Nut and bolt profiles.

Similarly, a radius for the root rounding in the internal thread could be calculated and is half of the bolt root value.

$$\text{Nut Root Radius} = \frac{\sqrt{3}}{24}P$$

Of course, if you add such radii to the nominal profile on a CAD model, there's a good chance your system will do the calculations for you.

Table 4.1 shows the common sizes for threads **M1 to M15**. Note that these are set into three orders of preference of use. Those in the first-choice column are usually more commonly available as bolts/nuts from suppliers than second- or third-choice sizes. For a full list of sizes and associated data, refer to ISO 261 or to a supplier's website.

TABLE 4.1
Sizes for 'M' Series Threads from 1 to 15 mm

Nominal Diameter				
First Choice	Second Choice	Third Choice	Coarse Pitch	Available Fine Pitches
1			0,25	0,2
	1,1		0,25	0,2
1,2			0,25	0,2
	1,4		0,3	02
1,6			0,35	0,2
	1,8		0,35	0,2
2			0,4	0,25
	2,2		0,45	0,25
2,5			0,45	0,35
3			0,5	0,35
	3,5		0,6	0,35
4			0,7	0,5
	4,5		0,75	0,5
5			0,8	0,5
		5,5		0,5
6			1	0,75
	7		1	0,75
8			1,25	1 0,75
		9	1,25	1 0,75
10			1,5	1,25 1 0,75
		11	1,5	1 0,75
12			1,75	1,5 1,25 1
	14		2	1,5 1,25[a] 1
		15	2	1,5 1

[a] Note: the size M14 × 1,25 is reserved for spark plugs only.

4.3.1 Tolerancing

Like the tolerancing described in the previous chapter, the specification of threads can make use of the system of **tolerance grades and classes** based on **limits and fits** and manufacturers often quote these on their technical datasheets for threaded components. Where manufacturers quote tolerances on threads in numerical form, these are often based on values obtained from standard limits and fits. Metallic threads do not generally make use of transition or interference fits (and where they do they are covered by separate standards) so **clearance fits** are the norm.

Assuming a roughly equal distribution of tolerances for both internal and external threads for clearance fits means that normally all the tolerances will be unilateral; the nuts will be larger and the bolts will be smaller than the diameters of the nominal thread profile. Particular care has to be taken however not to allow overly large tolerances on threads because large clearances can significantly compromise the contact area between the flanks of the threads and hence weaken couplings. Since threads have more than one important diameter (e.g. pitch, crest) we can assign tolerances to each, so threads often show two tolerance grade/class values.

In practice, and using the ISO 965–1:2013+A1:2021 standard, the tolerance grades (the widths of the tolerances) for internal threads are limited to grades 4, 5, 6, 7 or 8 for both the minor and pitch diameters while the tolerance positions are limited to 'G' or 'H'.

For external threads the tolerance grades are 4, 6 or 8 for the major diameter and 3, 4, 5, 6, 7, 8 or 9 for the pitch diameter. The tolerance positions used are 'e', 'f', 'g' or 'h'. This is shown graphically in Figure 4.3 using the nomenclature from the ISO standards.

EI = the fundamental deviation (minimum clearance) of the nut from the nominal profile
Ti = the tolerance range of the nut (internal) profile
ES = the maximum deviation of the nut profile from the nominal profile
and ES = EI + Ti

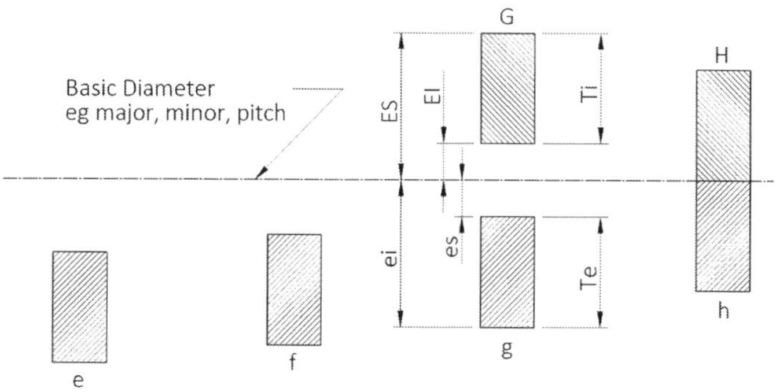

FIGURE 4.3 Tolerance grades and classes (see Chapter 3) for threads used in ISO 965-1.

and EI > 0 for position 'G' and EI = 0 for position 'H'
Correspondingly,
es = the fundamental deviation (minimum clearance) of the bolt from the nominal profile
Te = the tolerance range of the bolt (external) profile
ei = the maximum deviation of the bolt profile from the nominal profile
and ei = es—Te
and es < 0 for positions 'e', 'f', and 'g' and es = 0 for position 'h'

These positions and grades give a suitable range of clearance values to allow easy fitting of the threaded components and an allowable distribution of flank contact areas to ensure appropriate strength characteristics.

Table 4.2 shows a selection of the fundamental deviations (EI and es) for selected internal thread (G and H) and external thread (f and g) positions.

Tables 4.3 to 4.6 show a small selection of the tolerance range (T) values given in ISO 965–1 for the various internal and external threads and for minor, pitch and major diameters.

TABLE 4.2
Selected Fundamental Deviations for Internal and External Threads

Pitch P (mm)	Position G EI (μm)	Position H EI (μm)	Position f ES (μm)	Position g ES (μm)
0,5	+20	0	−36	−20
0,6	+21	0	−36	−21
0,7	+22	0	−38	−22
0,75	+22	0	−38	−22
0,8	+24	0	−38	−24
1	+26	0	−40	−26
1,25	+28	0	−42	−28
1,5	+32	0	−46	−32
1,75	+34	0	−48	−34
2	+38	0	−52	−38

TABLE 4.3
Selected Minor Diameter Tolerance Ranges for Internal Threads

Pitch P (mm)	Grade 6 Ti (μm)
0,75	190
0,8	200
1	236

TABLE 4.4
Selected Major Diameter Tolerance Ranges for External Threads

Pitch P (mm)	Grade 6 Te (µm)
0,75	140
0,8	150
1	180

TABLE 4.5
Selected Pitch Diameter Tolerance Ranges for Internal Threads

Major Diameter D (mm)	Pitch P (mm)	Grade 6 Ti (µm)
Over 5,6 to 11,2	0,75	132
	1	150
	1,25	160
	1,5	180

TABLE 4.6
Selected Pitch Diameter Tolerance Ranges for External Threads

Major Diameter D (mm)	Pitch P (mm)	Grade 6 Te (µm)
Over 5,6 to 11,2	0,75	100
	1	112
	1,25	118
	1,5	132

Suppliers' catalogues and websites will often quote the various tolerances for threaded components in the form of classes made up of ISO alphabetic positions and numerical grades or simply with their tolerance values (often translated from the position/grade tables in the standards).

Where a specification for an internal thread is given with a single tolerance class such as 'M10–6H', the class refers to both the pitch and minor diameters and where two classes are shown such as 'M10–5H 6H', then these relate to the pitch diameter and the minor diameter, respectively.

Similarly, where a specification for an external thread is given with a single tolerance class such as 'M10–6g', the class refers to both the pitch and major diameters and where two classes are shown such as 'M10–5g 6g', then these relate to the pitch diameter and the major diameter, respectively.

4.3.1.1 Example—M6 × 1 6H/6g

A fastener arrangement specified as '**M6 × 1–6H/6g**' will have a nominal (or 'basic major') diameter of 6 mm and a pitch of 1 mm. For the nut we can calculate (refer to Figure 4.1) or find from the standards tables the nominal pitch diameter and the minor (internal thread crest) diameters:

$$\text{Pitch Diameter} = \text{Nominal Diameter} - \frac{3}{8}Hx2 = 6 - \frac{3}{8}\sqrt{3}P = 5,350\,\text{mm}$$

and

$$\text{Minor Diameter} = \text{Nominal Diameter} - \frac{5}{8}Hx2 = 6 - \frac{5}{8}\sqrt{3}P = 4,917\,\text{mm}$$

From the tables of values in ISO 965–1 (or see the following text), we can find the 6H tolerance values (EI and ES) for the pitch and minor diameter of a nut using the known major (nominal) diameter (6 mm) and pitch (1 mm). These are:

EI = 0 (see Table 4.2), ES = EI + Ti = 0 +150 (see Table 4.5) = +150 µm for the pitch diameter and
EI = 0 (see Table 4.2), ES = EI + Ti = 0 + 236 (see Table 4.3) = +236 µm for the crest or minor diameter.

Applying the tolerances gives:

$$\text{Pitch Diameter}_{\text{nut}} = 5,350\,\text{mm}\,{}^{+150\,\mu m}_{+0\,\mu m}$$

and

$$\text{Minor Diameter}_{\text{nut}} = 4,917\,\text{mm}\,{}^{+236\,\mu m}_{+0\,\mu m}$$

For the bolt we find the 6g tolerances are:

es = −26 µm (see Table 4.2) and ei = es − Te = −26 − 112 (see Table 4.6) = −138 µm for the pitch diameter
es = −26 µm (see Table 4.2) and ei = es − Te = −26 − 180 (see Table 4.4) = −206 µm for the crest or major diameter.

This gives:

$$\text{Pitch Diameter}_{\text{bolt}} = 5,350\,\text{mm}\,{}^{-26\,\mu m}_{-138\,\mu m}$$

and

$$\text{Major Diameter}_{\text{bolt}} = 6\,\text{mm}\,{}^{-26\,\mu m}_{-206\,\mu m}$$

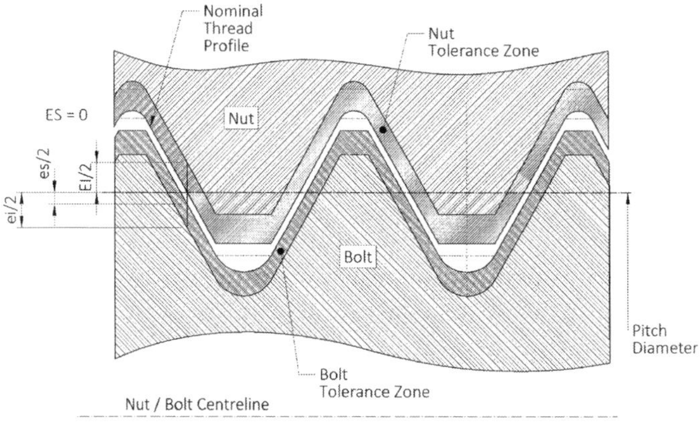

FIGURE 4.4 Nut and bolt (axially aligned) with tolerance indicators (not to scale).

Figure 4.4 shows a cross-section of the nut/bolt assembly and the tolerance zones that result from the previous calculations, assuming the bolt is axially aligned with the nut. Since ES = 0, there is no gap between the nut and the nominal thread profile when the tolerance lands at its lowest magnitude (when, additionally, EI = 0). This represents the situation where the nut would have its largest possible mass, its MMC. In this assumed extreme state the nut profile is considered to have perfect form (shape). There is a small gap (shown unhatched in white) that represents the minimum gap possible when the bolt is in its maximum material condition (es = –26 μm).

4.3.2 Thread Engagement

An important further consideration when specifying a thread may depend on the likely length of thread engagement. This is the length of bolt/nut interaction or how long the axes of the mating threads overlap. Clearly, longer engagement lengths will increase the flank contact areas and may allow for looser tolerances and larger clearances. However, over-long engagement is not cost effective and leads to the need to apply higher-than-necessary torque to secure an assembly. As an oft-used rule of thumb recommended engagement lengths vary from 1 to 1,5× nominal diameter for steel, 1,5–2× nominal diameter for cast iron or brass and around 2–2,5 for aluminium. Plastic threads are difficult to predict but often require quite long engagement lengths. Where thread engagement is important, it may be necessary to calculate joint strengths from design data for specific materials, possibly using finite element analysis. Manufacturers and suppliers often quote their own recommendations for specific materials. ISO 965–1 has advice on classifying engagement in terms of short, normal and long; for example, for an 'M6 × 1' thread, engagement lengths up to less than 3 mm are 'short', 3 mm up to less than 9 mm are 'normal' and 9 mm and above, 'long'. If necessary the thread engagement should be appended to the thread specification with the relevant letters S, M or L: for example,

$$\mathbf{M6 - 7H / 7g\ 6g - L},$$

for an M6 thread with internal thread class 7H for pitch and minor diameter tolerances and external classes 7g and 6g for pitch and major diameter tolerances, respectively, and a long engagement length.

One general consideration for thread specification is to address the quality of the thread manufacture. ISO 965–1 uses a very broad classification of 'fine' (not to be confused with the fine thread series!) for high precision threads, 'medium' for general use and 'coarse' (not to be confused with the coarse thread series!) for hot rolled or long blind holes. These standards give recommendations for the preferred tolerance grades/classes for the various combinations of tolerance quality and thread engagement. For example, for 'medium' quality threads with 'normal' engagement lengths, 6H is preferred for internal threads and 6g for external threads.

4.3.3 Chirality or 'Handedness'

All of the aforementioned threads, in fact, any thread in theory, can be either **right-handed** or **left-handed**. Right-hand threads are normally used in most applications, and the nut or bolt is generally tightened onto its mating part by applying torque clockwise looking in the direction of positive axial travel; in other words, when the nut or bolt is turned clockwise, it will travel away from the operator.

In applications where the predominant rotation of the fastener tends to loosen the thread, it is often necessary to use a left-handed thread so that the predominant rotation of the part will tend to tighten the coupling. A very common example of this is the left-sided (or port-sided) pedal on a bike. Looking in the direction of the pedal being in front of the crank, the pedal is tightened by applying the pedal spanner counter-clockwise and loosened by turning the spanner clockwise. Many pedals have threaded shafts which can also be accessed from the opposite side of the crank by means of an Allen key. Viewing with the crank in front of the pedal everything is reversed again, where loosening is achieved by turning counter-clockwise. This is a common source of confusion for many cyclists.

Specification of the **chirality** of a thread can be done by adding '**RH**' or '**LH**' to the thread description, for example, 'M6x1–6g LH'. Where there is no indication, 'RH' is assumed.

4.3.4 Multi-Start Threads

So far the coverage of threads has related only to single starts, that is, those that have a single helix. An important concept in specifying threads relates to what is called the '**Lead**'. This is the distance that a threaded component will travel when one full turn is applied to it in its assembly. When the number of thread **starts** is one, then Lead = Pitch. However, threads can also have 2, 3 or 4 starts commonly, and these will affect the lead as follows;

$$\text{Lead} = \text{Pitch} \times \text{No.of starts}$$

The starts are distributed evenly around the thread diameters so a two-start will have its starts at 180° to each other and a four-start at 90°. The thread helices run

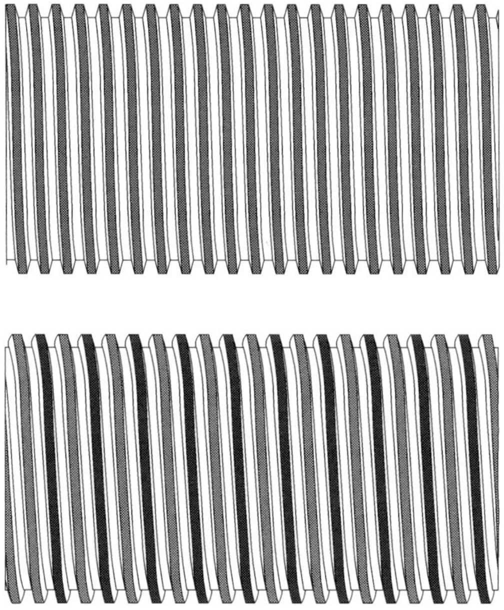

FIGURE 4.5 Single- and double-start threads.

parallel to each other up the body of the component, and this can result in greater helical angles than those on single-start threads. Figure 4.5 shows a simple **single-start** right-hand thread (above) and a simple **double-start** thread (illustrated with different shades, below), with the steeper helical angle apparent.

The advantages of using **multi-start** threads are that they allow a greater travel distance for a given rotation of a thread. This can be useful for positioning drives or for use on plastic bottle threads where there is a desire to quickly lock the cap down on a bottle with a single turn of the hand.

The disadvantage of multi-start threads is that they may be harder to manufacture. It can also be problematic that they do not tend to **self-lock**, although this is sometimes desirable depending on the design intent. The various 'M' series threads discussed earlier will self-lock; in other words, if a bolt/nut is subject to force along its axis (pulled), it will not tend to rotate and move along its axis. Such conditions, however, must also be considered in the context of friction, vibration and the use of possible locking devices.

4.4 UNIFIED THREADS

Unified threads are currently designated by the **American National Standards Institute (ANSI)** and the **American Society of Mechanical Engineers (ASME)**. They represent a group of threads that are derived in imperial (Inch) units although ISO recognize these in both inch and equivalent metric sizes. The basic form of unified screw threads is the same as the ISO 'M' series. They are based on a thread

TABLE 4.7
ANSI Unified Thread Designations

Type	Descriptions
UN	Constant pitch
UNC	Coarse pitch
UNF	Fine pitch
UNEF	Extra fine pitch
UNS	Special diameter
UNJ	Constant pitch with rounded root radius RRR = 0,15011 to 0,18042P
UNJF	Coarse pitch RRR = 0,15011 to 0,18042P
UNJEF	Fine pitch RRR = 0,15011 to 0,18042P
UNRE	Extra fine pitch RRR = 0,15011 to 0,18042P
UNR	Constant pitch RRR => 0,108P
UNRC	Coarse pitch RRR => 0,108P
UNRF	Fine pitch RRR => 0,108P
UNREF	Extra fine pitch RRR => 0,108P

angle of 60° and have the same proportions for depth, crest and root locations and also allow for the rounding of the root profile.

A typical simple thread designation includes the nominal diameter, the number of threads per inch and the thread type, for example, '**½-20 UNF**'. However, there are many thread types covered by the UN inch designations as listed in Table 4.7. Details of all of these are given in the ANSI B1.1 and B1.15 standards.

4.5 WHITWORTH THREADS

Although these were introduced in the 1800s, these threads are still frequently found today. The basic profile of the Whitworth thread has a thread angle of 55° and rounded root and crest areas as shown in Figure 4.6.

Thread types for common Whitworth profiles are shown in Table 4.8.

These threads were originally defined in inch-based fractional sizes, but metric-converted sizes are also available, although, like metric versions of unified threads, the sizes carry several decimal places. Care has to be taken using such conversions because they involve either rounding the various values or specifying very precise sizes and implying high manufacturing precision. This can be ameliorated by judicious and clear tolerancing to show the desired design intent.

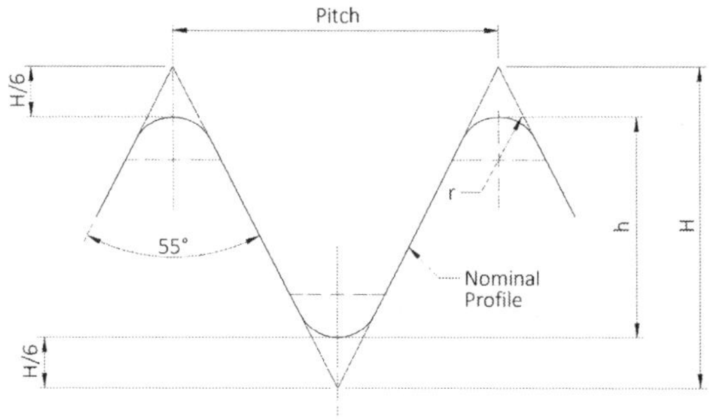

FIGURE 4.6 Whitworth geometry.

TABLE 4.8
Whitworth Thread Designations

Type	Descriptions
BSW	British Standard Whitworth thread
BSF	British Standard Fine thread
BSP	British Standard Pipe thread

4.6 BA AND SMALL THREADS

The British Association thread type is now obsolete; however, it is a small (<0,25 in) designation regularly found in older designs and especially in electronics assemblies, for example, in hifi and scientific equipment.

For miniature threads in the ISO system (inc watch threads), the BA series has been superseded by metric threads described in the **ISO/R 1501** standard, which covers diameters from 0,3 to 1,4 mm. The thread designation is given the prefix 'S', for example:

$$\mathbf{S-0.6\times0.15-4\ H6}$$

which shows the prefix, diameter × pitch, the nut pitch grade and the nut tolerance grade.

Note that the prefix 'S' is also used in the designation of 'straight' as opposed to 'tapered' on pipe threads.

4.7 PIPE THREADS

There are very many types of pipe threads in use around the world, largely because they cover a wide range of industries, and there are very many historical fittings that need attachments. Important standards are found in **ISO, ANSI, JIS, SAE, BS** and **DIN** standards, for example. For these reasons the various pipe threads appear across many borders, and it is useful to have some overall familiarity with various types. In Europe and many commonwealth countries, perhaps the most

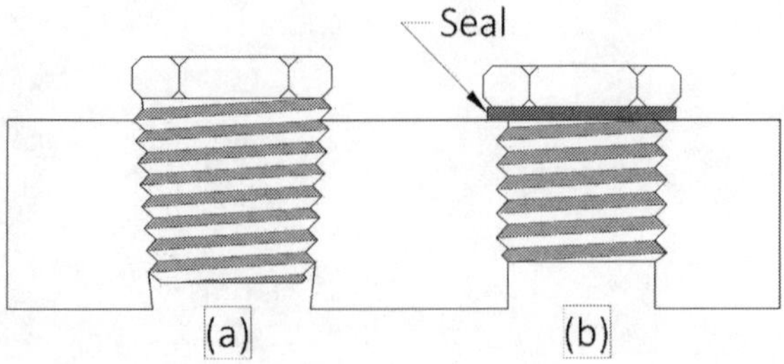

FIGURE 4.7 Pipe thread example profiles.

common pipe threads are the **ISO 7–1:1994** and **ISO 228–1:2000**, which were derived from earlier BS threads, for example, BS 21: 1985. These British Standard Pipe (BSP) and British Standard Pipe Tapered (BSPT) standards were in turn based on earlier versions of the BS 21 standard. The current versions give metric equivalent sizes for the original imperial inch sizes on which the standards were based. They are also formed on a standard basic Whitworth thread profile.

In the US, the predominant current standards are known as the **National Pipe threads (NPT)** and **National Standard Fuel Type threads (NPTF)**.

In the far-east the JIS pipe standards are very common. The basic JIS profiles are very similar to the ISO/BS types and in some cases interchangeable.

The SAE standard of pipe thread was developed for use by the automotive industry and is commonly used there and in some related product areas.

Two main types of pipe thread are in common use and represent two different ways of achieving a seal on the connections of fluid-carrying pipes and hose connectors. The first type is a **tapered thread** design as shown in Figure 4.7(a). When the threaded components are tightened against each other, the threads deform and form a seal along the body of the thread, sometimes helped by the addition of sealing tape or compound applied to the male part prior to assembly. The female part of such an assembly may or may not be tapered, but the male part always is. The second type of seal is achieved by adding a seal such as a rubber washer to the collar of the male component, as shown in Figure 4.7(b), or to the bottom of the female thread.

4.8 SQUARE, ISO TRAPEZOIDAL, ACME AND BUTTRESS THREADS

Square, trapezoidal, acme and **buttress** threads are normally used for power transmission and positioning applications and can commonly be used in single- or multi-start configurations.

4.8.1 Square Threads

Square threads give an efficient means of transmitting power to move mechanical parts like tool posts on machines such as lathes or on heavy-duty lifting devices such as car jacks. They are often operated by means of simple handwheels or

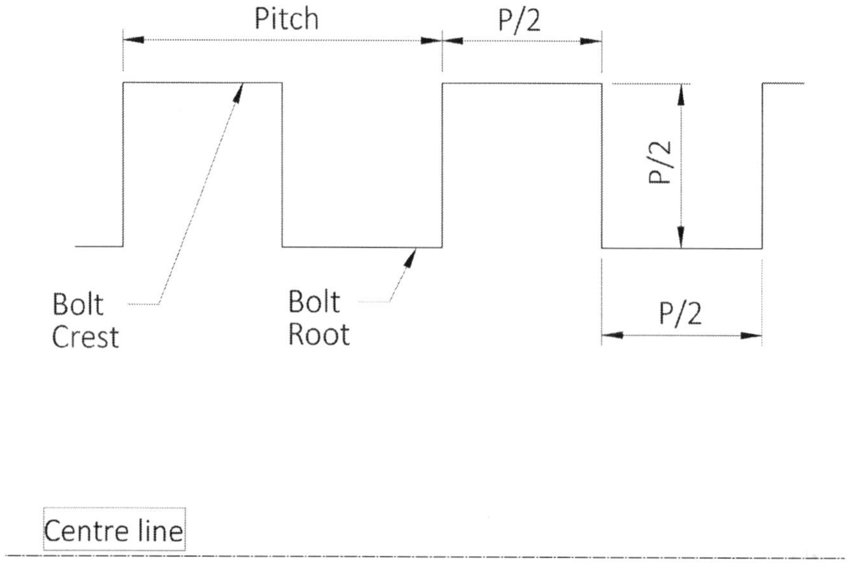

FIGURE 4.8 Square thread profile (bolt).

cranks. As the name indicates, they can have a simple square cross-section as shown in Figure 4.8. Because this thread form can be difficult to manufacture and is normally made for a very specific application, it is usually made to order and a range of diameter and pitch values can be chosen.

Square threads can have a wide variety of diameter-to-pitch ratios, but 5 to 8 is typical for sizes up to a diameter of 100 mm. A typical designation of a square thread is as follows:

$$\mathbf{Sq\ 30 \times 12\ (P6)},$$

where nominal diameter = 30, lead = 12 mm and pitch = 6 mm, and therefore it is a double (12/6)-start thread.

When a single-start thread is used, a simpler 'Sq 30 × 6' designation suffices.

In practice square threads usually have some small radii at the root and a small chamfer at the crest of the threads, and tolerances are added to ensure reasonable fits.

4.8.2 Trapezoidal Threads

Trapezoidal threads are also used to transmit power although they are less efficient than square forms, but they are stronger and generally easier to manufacture. The basic form of an ISO trapezoidal thread is shown in Figure 4.9.

The main range of these threads is described in **ISO 2901: 2016, ISO 2902: 2016** and **ISO 2904: 2020** with tolerance data in ISO 2903: 2016.

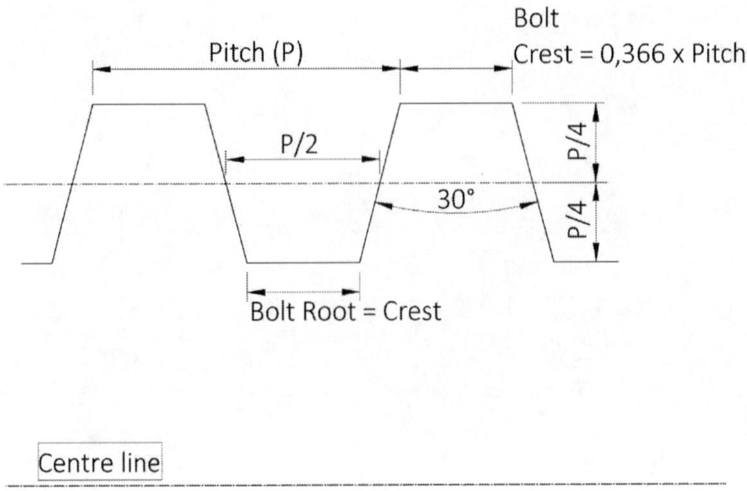

FIGURE 4.9 Trapezoidal thread form (ISO).

A typical designation for these threads will have the prefix 'Tr', for example,

$$\mathbf{Tr\ 60 \times 18\ (P9)\ LH},$$

where nominal diameter = 60 mm, lead = 18 and pitch = 9 (indicating a double (18/9)-start left-handed thread).

4.8.3 Acme Threads

Acme threads also have a trapezoidal type of cross-section and are common in US equipment. They have a smaller draft angle than the ISO threads, as shown in Figure 4.10.

As with the ISO trapezoidal forms, radii, chamfers and tolerances can be added to the profiles of both nut and bolt to aid assembly and operation. The specification of acme threads is given in **ANSI B1.5**.

4.8.4 Buttress Threads

Buttress threads are commonly used to apply a holding force in one axial direction. For this reason they have an asymmetric profile as shown in Figure 4.11.

Buttress threads are often made to order for a particular application, and they are used in a variety of configurations. They are characterized by having a steep pressure or leading face, typically lying at 0° to 10° to the vertical and a back or trailing face with a much shallower angle (normally 45°) to provide strength.

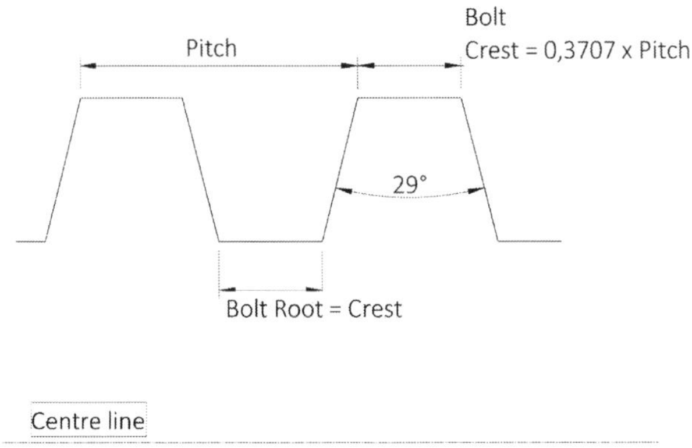

FIGURE 4.10 Acme thread form.

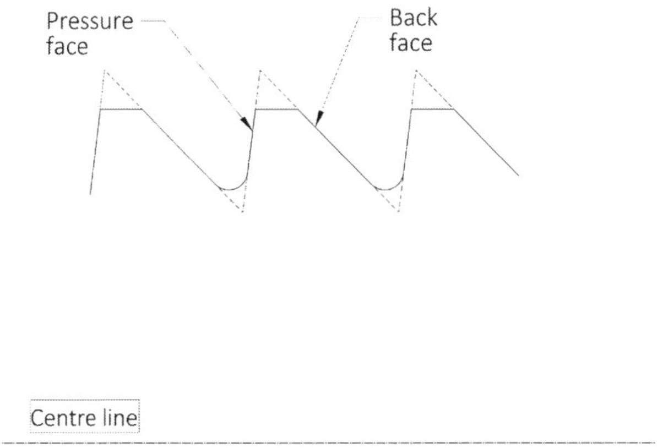

FIGURE 4.11 Buttress thread form.

4.9 PLASTIC THREADS

The thread types discussed previously were developed primarily for use with metals in mind, and although **plastic threads** are also made in some of these forms, they are likely to require quite different engagement lengths and may be made to different tolerance classes and grades. The material characteristics of most plastics mean that they are generally not well-suited to fine thread configurations, so even when they are made in standard (e.g. M8) sizes, they tend to be made with coarse-graded thread types.

However, many plastic threads are made to interface with metallic components because they have relatively good sealing characteristics for very little cost. They are particularly useful, for example, as protective seals on the ends of newly manufactured pipes, and these are therefore made with relatively fine pipe type threads.

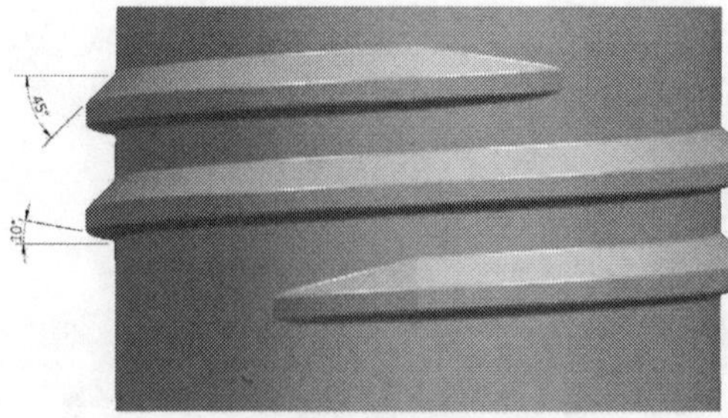

FIGURE 4.12 Plastic thread form example for liquid storage.

There are many standards that relate to the numerous types of plastic threads, and these should be sought for a particular application, for example, for plastic pipe threads.

One very common application of the use of plastic-to-plastic threads is in the sealing of plastic containers, most numerously in drinks bottles but also for other consumer products such as liquid cleaners. Figure 4.12 shows one thread type that is used for liquid storage, in this case for the outside of the bottle neck.

Note that the thread profile is similar to a buttress thread; in this case, the shallow pressure face is at 10° to resist any pressure forces from the fluids that will act to try to push the cap from the bottle. The **DIN 6063** is commonly used for the specification of plastic container threads in ISO regions.

4.10 DRAFTING CONVENTIONS

Most commonly, if standard thread types are being used, it is sufficient only to give a simplified view of a thread and attach the necessary details by means of dimensions and callouts or with tables of details for parameterized parts and parts with, for example, multiple threaded holes. (Modern CAD systems may also use full thread geometry if a part is to be made using additive manufacturing equipment.) Some simplified examples follow.

4.10.1 Male Threads

Figure 4.13 shows a simplified view of a component with a callout for one of the threads.

Most 3D feature-based CAD systems will recognize the properties of threads if they were constructed as standard threads in the 3D modelling stage, and several parameters may be available to build more elaborate callouts.

4.10.2 Female Threads

In a similar fashion, internal thread callouts can be created on 2D drawings or 3D PMI depictions when these have been entered at the modelling stage. Figure 4.14 shows internal thread parameters for simplified views of holes.

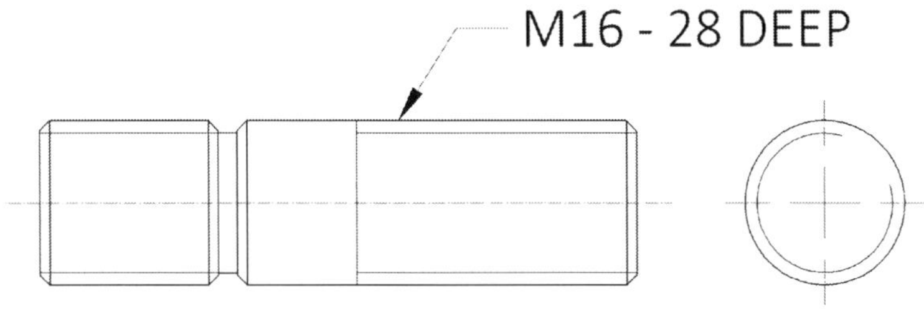

FIGURE 4.13 Simplified male thread with a callout.

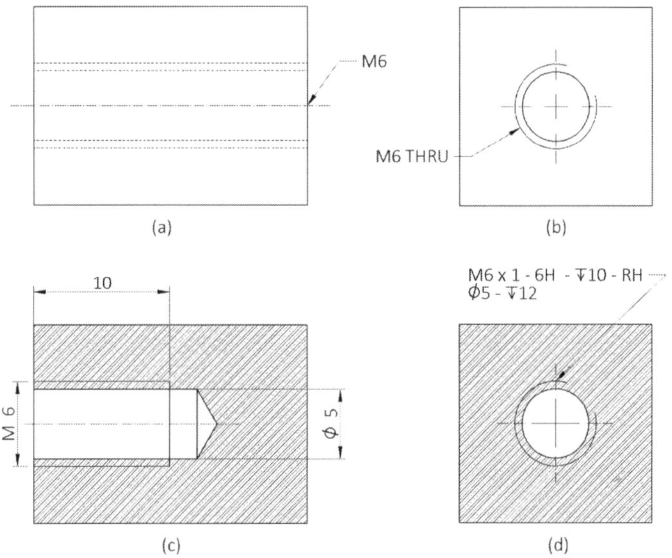

FIGURE 4.14 Simplified female (internal) threads.

4.11 SPLINES

As we'll discover in Chapter 10 there are two meanings of the word '**spline**' that are used in engineering drawing. The first, described here, relates to a feature comprising longitudinal protrusions in a pattern around a shaft or hole feature. Spline features, like threads, usually consist of a matching pair of shaft and hole that form a joint, which is commonly used to transmit power, for example, on drive shafts or steering column attachments. Traditionally these would be time consuming to draw, and so, like with threaded features, a simplified convention was often used whereby only a small proportion of the feature was shown on a drawing. With modern CAD systems, it is relatively easy to display an entire spline feature. Nonetheless, many engineers prefer a simplified view as it can make a drawing or PMI depiction clearer and less cluttered looking. In Chapter 1 (Figure 1.10) we saw an end elevation simplified view of a splined shaft. This method, appropriately dimensioned, can be used for any cross-sectional shape of protrusion. There are, however, standard configurations of spline joints, namely square-sided (**ISO 14:1982**) and involute (**ISO 4156-1:2021**), and the methods for depicting these are given in **ISO 6413:2018**. Also, as with threads, the system of limits and fits can be applied to detail standard tolerance values. A simple example will demonstrate a simplified but full specification for a standard spline as shown in Figure 4.15.

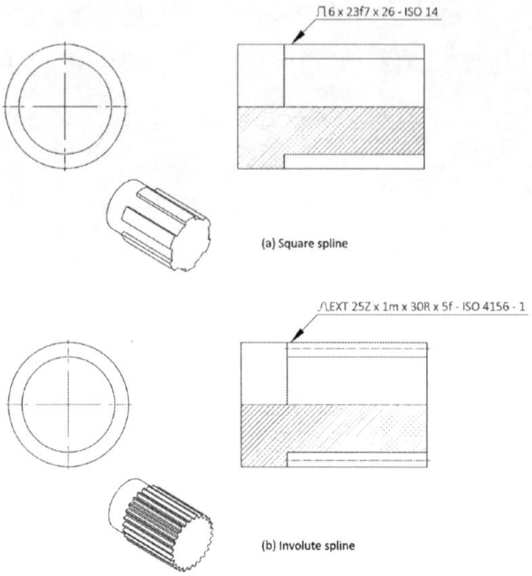

FIGURE 4.15 Splined shafts (not to scale).

Figure 4.15(a) shows a square profiled spline with a callout indicating, number of teeth = 6, minor diameter = 23 mm with tolerance class = f7 (note the position grade reference is reversed in some older standards), major diameter = 26 mm and design standard = ISO 14.

Figure 4.15(b) shows an involute profiled spline with a callout indicating spline type = external feature (shaft-based), number of teeth = 25, module number = 1, pitch diameter = module number × number of teeth = 1 × 25 = 25 mm, pressure angle = 30°, tolerance class = 5f, design standard = ISO 4156–1; the major and minor diameters are dependent on calculations involving the tolerance values as set out in the standard in ISO 4156–1.

The same methods for callouts are used on internal spline features.

4.12 KNURLING

Knurling is normally carried out with a specialized tool with standard dimensions and therefore does not require specific detail, although this can be added if necessary. Simplified methods of depicting straight and diamond pattern knurling are shown in Figure 4.16.

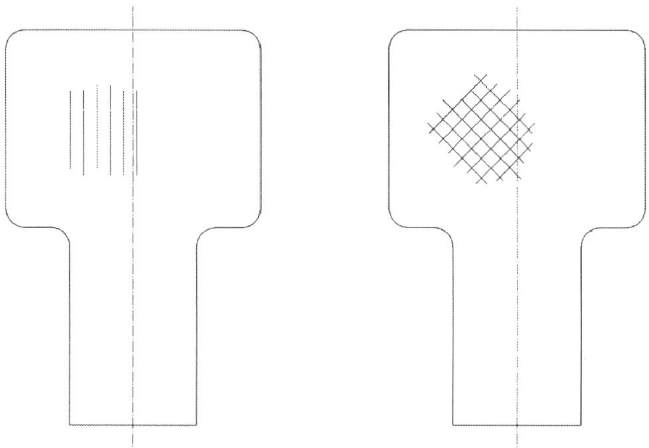

FIGURE 4.16 Straight knurling (left) and diamond knurling (right).

4.13 COMMENTS

The introduction to methods of depicting threads, splines and knurled features given here is not exhaustive but gives some instruction on how to use the data given in the relevant standards. There are many sources of data available on the Internet as well as in the standards. While most CAD systems have built-in commands to make the depiction of threads relatively easy, most do not support spline features explicitly. Some CAD systems do have support for gear assembly calculations, which can be used to create spline components, including helical splines.

Now that we have looked at typical standards-based methods for drawing, dimensioning and tolerancing some common engineering features, we can turn to the more advanced topics necessary to facilitate geometrical tolerancing.

5 Introduction to Geometric Tolerancing and Form Tolerances

5.1 KEY CONCEPT—FEATURES

Geometric dimensioning and tolerancing as implemented by ISO and ASME are **symbolic languages** that are used to fully define the features of a part in terms of their size, **form** (shape), **location** and **orientation**. The advantage of a symbolic language is partly down to the fact that it is somewhat independent of specific natural languages such as the different versions of English or any other language. The disadvantage of a limited symbolic language is that it cannot be used to describe every possible design/manufacture/inspection scenario so there is often a need for augmented **notes for clarification purposes**. Key concepts in understanding the use of geometric dimensioning and tolerancing include **datums**, **definitions of the different types of feature** and the use of **tolerance indicators**. This follows on from the treatment in earlier chapters describing basic dimensioning and tolerancing and the concept of material conditions (**Chapter 3**). In the ISO system many of the concepts and the language used to describe the different types of feature on a component are outlined in **ISO 14405** and **ISO 1101:2017**.

The word 'feature' takes on many meanings in the analysis of manufactured parts and perhaps nowhere more so than in the standards that describe **Technical Product Specification (TPS)** and **Geometric Product Specification (GPS)**. It is important to understand the various types of feature and to understand that a given feature can be of several types at once. Think of a feature as a noun and the type as an adjective that describes the feature's role or attributes.

Let's look at some of the different types of features that are important. We can begin by observing that GPS is involved with depicting designs in the context of their nominal form with their allowed deviations and subsequently with the actual manufactured form and its validation through metrology.

5.1.1 Internal versus External Features

As a reminder, **internal features** are spatial features with internal faces, for example, slots and holes, whereas **external features** are those with external faces such as shafts or rectangular blocks.

 DOI: 10.1201/9781003395072-5

5.1.2 Features of Size

A major key concept in GPS is the definition of a **feature of size**. In essence, a feature of size is any feature that can be characterized by a measurement across its expanse between two opposing points. Figure 5.1 shows some simple examples of what should and what generally should not be regarded as a feature of size according to ISO.

Common features of size include two parallel opposed planes, cylinders, spheres, cones or torii. These can all be internal or external features.

5.1.3 Integral versus Derived Features

Integral features are features that consist of, or represent, surfaces on a part and may or may not be features of size. Conversely, a **derived feature** is one that depicts a theoretical point, line or plane that is constructed, or derived, from a feature of size. Typical examples include an axis of rotation of a cylinder or cone, a median line or a medial surface, for example, in a slot. These features might be

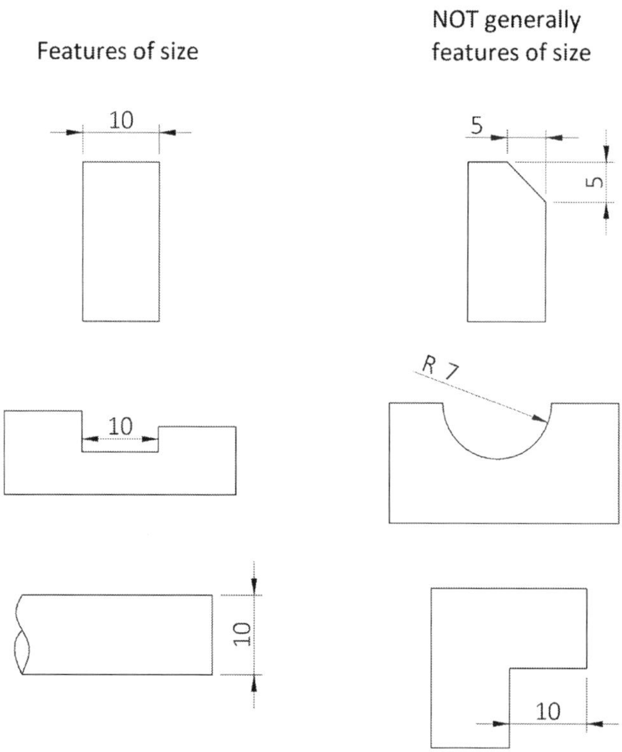

FIGURE 5.1 Features of size.

shown on drawings but to measure to or from them requires the use of auxiliary equipment that can simulate say a median plane between two faces of a slot.

5.1.4 Nominal, Real, Extracted and Associated Features

This differentiation of feature types relates to the design, manufacturing and validation phases of a part. When we create 3D models of parts in CAD systems we are making a model that represents reality in the context of a particular purpose (e.g. to enable manufacture). In the world of GPS standards, these models are considered to be models of the nominal geometry, an idealized form with exact sizes that may be accompanied by attribute information such as permitted size deviations (tolerances). The modelled features contained in such models are therefore described as **nominal features**.

When a part is manufactured it will of course not be of perfect form and therefore not conform exactly to the modelled part. The physical features are therefore referred to in the ISO 14405 and ISO 1101:2017 standards as **real features**. They will not have perfect shape or be oriented exactly right.

If a real part's features are measured in some way, these measurements will form a kind of measurement model of the part and the measured features, are named **extracted features**. Note that the ISO standards imply more emphasis on the metrology aspects of geometric dimensioning than in the corresponding ASME 14.5 and related standards. The concept of extracted features therefore can imply the use of complex measuring equipment such as **measurement arms**, **coordinate measuring machines (CMMs)** or **laser scanning devices**; however, it can also include more traditional set-ups with **vee blocks**, **gauge pins**, **micrometers** or **callipers,** which many regard as implied by the ASME standards.

Once a set of measurements of a feature is recorded, the extracted feature may have an additional idealized model created from it. Such features may be created by fitting idealized surfaces to them using the techniques discussed in **Chapter 3**, for example, using least-squares or Chebshev (minimax) methods amongst others. The resulting idealized model is said to be associated with the real or extracted geometry and is therefore referred to as an **associated feature**.

A graphical sketch of the geometric feature types is shown in Figure 5.2.

5.2 MAXIMUM MATERIAL VIRTUAL CONDITION (MMVC) AND LEAST MATERIAL VIRTUAL CONDITION (LMVC)

In Chapter 3 we discussed the concept of maximum material condition (MMC) and least material condition (LMC). Note that in ISO standards when a feature is at its MMC or LMC, it is at its maximum or minimum size and the corresponding dimensional size is said to be its maximum material size (MMS) or least material size (LMS).

It is often desirable to specify that a geometric tolerance (GT) should be met at one of these conditions, for example, that a value of 'straightness' applies

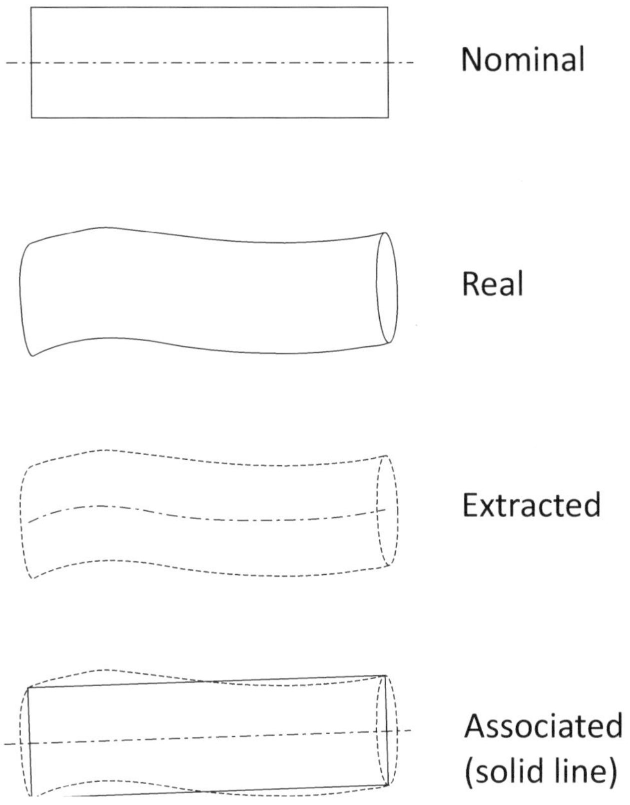

FIGURE 5.2 Geometric feature types.

when a feature is at its MMC. Specifying MMC on a feature is known as applying a maximum material requirement, or MMR, on a GT, and it applies only to features of size (a surface cannot have a size so cannot be at any specific material condition).

The effect of the MMR is to specify a GT as well as a dimensional tolerance and therefore to create what is known as a **maximum material virtual condition**, or **MMVC**.

For an external feature of size, for example, a shaft:

$$\mathrm{MMVC} = \mathrm{MMS} + \mathrm{GT}.$$

The MMVR effectively represents the largest allowable positive material deviation of a feature from its nominal or modelled form.

Therefore, for an internal feature of size, for example, a hole:

$$\mathrm{MMVC} = \mathrm{MMS} - \mathrm{GT}.$$

As you would expect these relationships differ for the least material requirement (LMR), which specifies a GT when a feature is realized at its LMC. This gives rise to a **least material virtual conditionor LMVC**.

For an external feature of size, for example, a shaft:

$$\text{LMVC} = \text{LMS} - \text{GT}.$$

For an internal feature of size, for example, a hole:

$$\text{LMVC} = \text{LMS} + \text{GT}.$$

The MMVC and LMVC are useful concepts when constraining datum features as we will see. Before commencing we should first look at some of the detail in defining what exactly GTs are.

5.3 GEOMETRIC METHODS OF DIMENSIONING, TOLERANCING AND VALIDATION

There are several other entities that are referred to as features in the ISO standards that we may come across later; however, for now we can use those discussed earlier while we introduce the feature-based geometric dimensioning, tolerancing and validation methods referred to in the ISO GPS standards.

The ISO 1101:2017 standard defines four tolerance types: form, orientation, location and run-out. The standard outlines the symbolic language that is used to define tolerances and how they are presented in the ISO GPS system. The symbols defined therein relate to the geometric constraints that can be applied to features within a part.

5.4 FORM TOLERANCES

Form tolerances relate to the shape of a feature and as such do not require reference to any datum system so they are largely self-contained. The set of constraints and their symbols are shown in Table 5.1.

TABLE 5.1
Form Tolerance Symbols

Symbol	Name
—	Straightness
⏥	Flatness
○	Roundness
⌭	Cylindricity
⌒	Profile any line
⌓	Profile any surface

These tolerances are used to control the form (or shape) of individual elements irrespective of the overall geometry of a part. Because they do not require reference to other features, they are commonly used to constrain datum surfaces as we shall see. They are also particularly useful for controlling features where the shape is crucial to the operation of a feature, for example, on cams.

Let's look at the basic use of the form symbols. To keep the illustrations as clear as possible, we will not show dimensions that are not relevant to the purpose of the examples.

5.4.1 Straightness

Figure 5.3 shows a basic **straightness** tolerance being applied to the surface of a shaft. The tolerance symbol and its range are shown in the rectangular **'tolerance indicator'** whose leader line points to the surface being constrained.

The value of 0,6 mm shown in the tolerance indicator means that the line shown must lie within two parallel bounding lines that are 0,6 mm apart, regardless of the actual diameter. Straightness tolerances can be applied to any outline that is considered a straight line and could therefore appear on a side view of a plane or a cylinder as in the example chosen in Figure 5.3. Such a tolerance might be validated in a workshop or metrology laboratory by inserting a cylindrical component into a vee block set-up and using a profile gauge to run along any longitudinal line on its surface and on a plane that passes through both the line and the central axis. Alternatively, CMM, measurement arm or a laser scanning device could be used to scan and, with appropriate analysis tools, validate compliance of the part/feature. Supporting software tools that help in the validation processes are becoming increasingly popular.

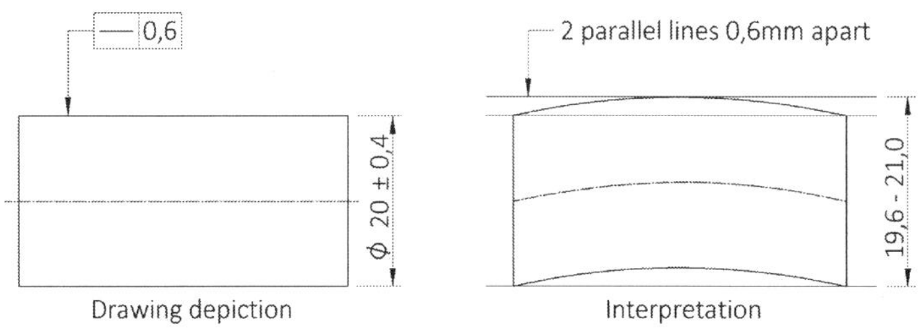

FIGURE 5.3 Straightness applied to a surface element.

Notes:

Because no modifiers are shown the principle of independence applies.

Since all two-point measurements must be within the size tolerance, then the minimum circumscribing envelope will occur at maximum diameter (20,4 mm) and maximum deviation (0,6 mm) from straight, giving 21,0 mm.

The maximum possible inscribing envelope size would occur at the minimum two-point diameter (19,6 mm) and perfect straightness (deviation = 0) giving 19,6 mm.

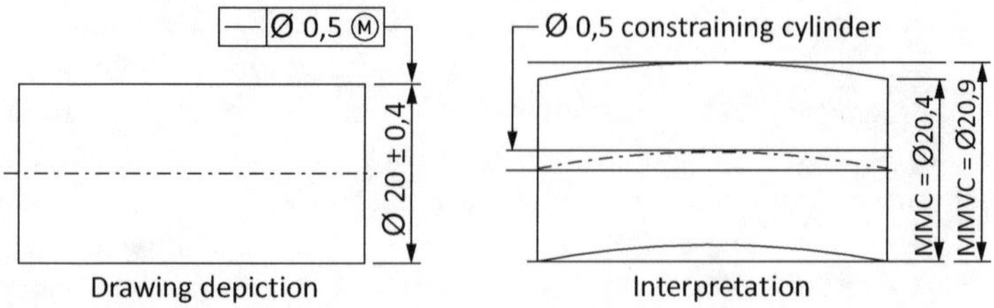

FIGURE 5.4 Straightness applied with MMR.

Notes:

The straightness tolerance indicator leader line points to the diameter dimension to show that it applies to that feature of size. The tolerance of 0,5 mm in the indicator is preceded by a diameter symbol and followed by the MMR symbol.

This indicates a straightness tolerance zone consisting of a constraining cylinder on the central axis of diameter 0,5.

The MMR gives rise to an MMVC.

For an external feature, MMVC = MMS + GT (straightness tolerance) = 20,4 + 0,5 = 20,9.

Straightness tolerances can be applied to features of size, and when this is done the tolerance zone applies to any straight line element of the feature. For example, if we apply a Straightness tolerance to a shaft it can be shown as seen in Figure 5.4.

The GT in this case (straightness) is applied to the cylindrical feature of size and the tolerance zone can be thought of as a cylinder of 0,5 mm diameter into which the feature's median line must fit.

5.4.2 Flatness

A flatness tolerance is similar to a straightness tolerance except that it applies to a complete surface or a designated part of a surface. In the example given in Figure 5.5 a height for a plate part is given and the envelope principle is invoked. This means that the feature, regardless of its deviations from flat must fit within the 20,4 mm envelope. The flatness tolerance zone then must fit within the envelope and cannot be added to the dimensional tolerance. Also, the size of the flatness tolerance must also be less than or equal to the dimensional tolerance since if the flatness tolerance was larger than the dimensional tolerance it would break the dimensional tolerance's criteria.

It is worth noting then that at MMC, when the feature is at its largest size there is no room for any deviation from flat. If the feature comes in at a slightly smaller size, say 20,3 mm, then an allowable deviation from flat of 0,1 mm is possible, and if the feature comes in at 20,2 mm then the full 0,2 mm flatness tolerance is possible. Further reductions in the dimensional tolerance cannot result in further allowances for deviation from flat, however, because the flatness tolerance stipulated will not allow a value greater than 0,2 mm.

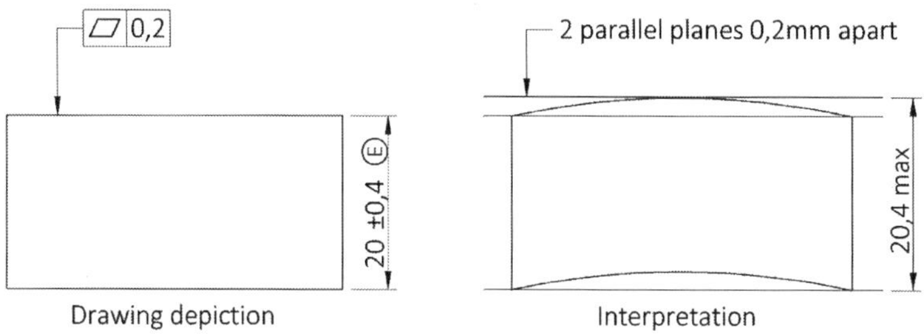

FIGURE 5.5 Flatness tolerance with envelope requirement modifier.

Notes:

Because the envelope modifier is shown the envelope requirement applies.

All two-point measurements must be within the size tolerance AND the feature must fit within an envelope no wider than the maximum toleranced value of 20,4 mm.

5.4.3 Roundness

An example of a depiction of a **roundness** tolerance being applied to a shaft is given in Figure 5.6.

Since the dimensional tolerance and the roundness tolerance are independent of each other, they can be validated separately. The constraining roundness tolerance zone circles are concentric with each other, but there is no constraint on their positions or diameters; however, the difference in their radii must be no greater than 0,2 mm. In theory the roundness tolerance could be greater than the dimensional tolerance since the values of m1, m2, m3, etc. do not need to share a common centre and variations in shape can accommodate this. In practice, however, it is not common that a roundness value greater than the dimensional variation is specified.

Practical validation of roundness can present a considerable challenge using basic workshop equipment. It is possible to get estimates of roundness with profile gauges and vee block set-ups, but metrology departments have historically also had roundness measuring machines for demanding work. These required substantial skills to set up and use. Increasingly, expensive CMMs and scanning devices are being used to more easily find centres and estimate deviations.

When roundness is constrained on a cylindrical surface, the tolerance zone lies on a plane normal to the cylindrical axis. If the feature being constrained is instead spherical, its roundness is measured in a plane where the sphere and circle's centre points are common.

When the roundness of an integral feature is not cylindrical or spherical, then the ISO GPS standards demand that a direction of measurement is specified. This is a practical consideration since using conventional measuring equipment (dial gauges or needle-based profiling devices) could make it difficult to make readings normal to an axis on shapes such as steep cones, for example. In most cases it is convenient to measure radii normal to the surface in question. Specifying this direction may make use of a datum feature, which will be covered in the following chapter.

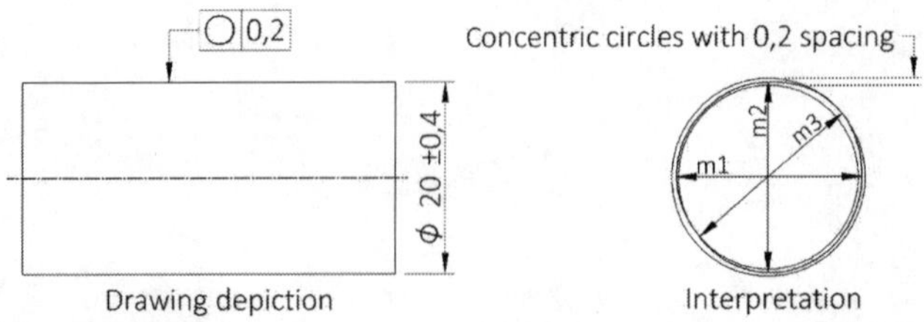

FIGURE 5.6 Roundness tolerance applied to a shaft.

Notes:

The roundness tolerance indicator leader line points to the cylinder, which indicates that it applies to any cross-section along the cylinder.

The principle of independence applies by default.

This indicates a roundness tolerance zone consisting of two concentric circles with an all-round spacing of 0,2 mm.

Independently, there is also a symmetric dimensional tolerance of +/-0,4 mm that must be met at any cross-section. Thus, any two-point measurement taken (m1, m2, m3, etc.) must be within 19,6 and 20,4 mm.

5.4.4 Cylindricity

A simple **cylindricity** tolerance is depicted in Figure 5.7.

5.4.5 Profiles

Profile tolerances were originally used for setting up tolerance zones around curved lines and surfaces. In earlier ISO standards their use for straight lines was discouraged; however, with the latest standards covering profile tolerances, specifically (**ISO 1660: 2017**), they are embraced and their use in practice is increasing.

Although the presentation in this section is for form tolerances, the use of profiles naturally raises issues around the desire to orientate and position profiles, and to achieve satisfactory means of doing this, we should define datums, which is the subject of the following chapter.

Also, in the coverage of profile tolerances for form, we will introduce some further features and **tolerance modifier symbols**.

Three simple line profile tolerance examples are shown in Figure 5.8(a–c). Note that profile tolerances can be applied to internal and external line features for both integral (i.e. feature outlines) and derived (e.g. centre lines) feature types.

In a manner similar to a straightness tolerance, the **line profile** tolerance defines a region around the profiled line, which is normal to its tangent and with a width value as specified. This is achieved by defining a rolling circle whose centre point travels along the specified line as shown in Figure 5.8(a).

In Figure 5.8(b) there are two modifier symbols introduced. Firstly the '**between**' symbol shows that the tolerance applies between the labels 'A' and 'B'. The '**CZ**' (**combined zone**) modifier specifies that each of the features between

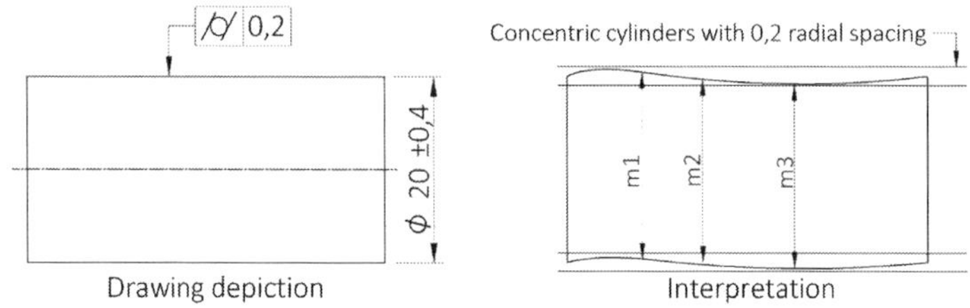

FIGURE 5.7 Cylindricity tolerance applied to a shaft.

Notes:

The cylindricity tolerance indicator leader line points to the cylinder, which indicates that it applies to the whole cylinder length.

The principle of independence applies by default.

This indicates a cylindricity tolerance zone consisting of two concentric cylinders with an all-round spacing of 0,2 mm. Their axis positions and diameter are unconstrained.

Independently, there is also a symmetric dimensional tolerance of 0,4 mm that must be met at any cross-section. Thus any two-point measurement taken (m1, m2, m3, etc.) must be within 19,6 and 20,4 mm.

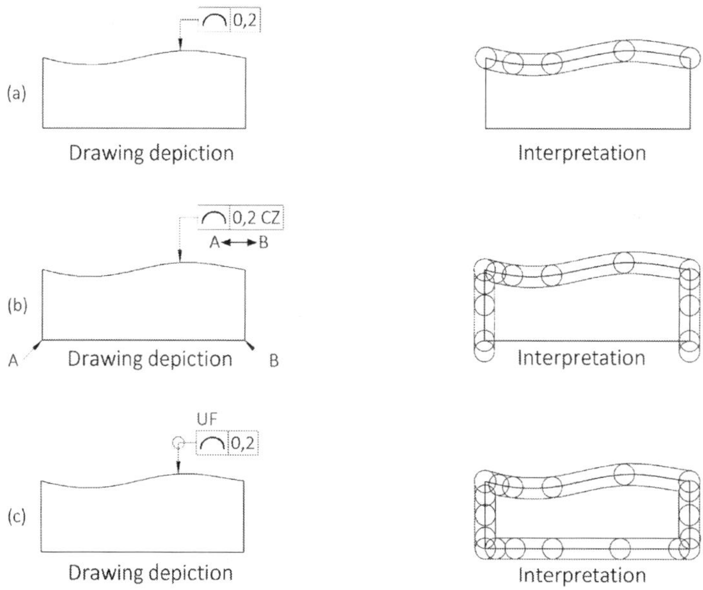

FIGURE 5.8 Profile tolerance on a line(s).

'A' and 'B' should have their individual tolerance zones defined and then combined. This can be a subtle and ambiguous depiction. For example, where the tolerance zones of the straight-line segments should extend linearly past the feature itself, the tolerance zone for the curve is not fully defined. The ISO standard states that in defining a curve using NURBS or other algorithms, the control or interpolation points should be positioned and the method of interpolation (the algorithm) should be specified. In practice this may imply that the means of extrapolating a curve beyond the feature end point may be necessary to complete the tolerance zone, and, if this detail is important, it too should be defined. In other words, the shape of the tolerance zones where the curves meet other features needs defining.

In Figure 5.8(c) two further modifier symbols are introduced. Firstly the '**UF**' (**unified feature**) modifier is specified, and this applies to a set of features making up a new unified feature. The '**all around**' symbol is added to the leader line to indicate that the new feature involves the whole profile. The 'UF' symbol means that the feature is unified first and then subsequently toleranced. The tolerance zone is therefore clear as the rolling circle centre point simply follows the nominal profile around the feature.

Surface profile tolerance examples are shown in Figure 5.9.

In its simplest form, a surface profile is very similar to a line profile. In the previous example, the tolerance zone exists across the whole 3D surface that has been extruded from a 2D profile.

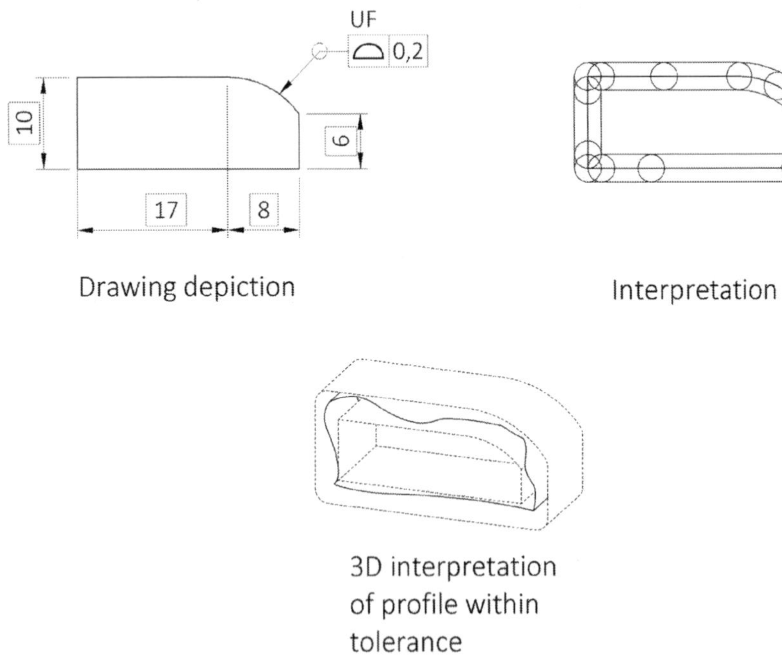

FIGURE 5.9 Surface profile form tolerance.

5.5 COMMENTS

The examples given in this chapter relate to the shapes, or 'form', of the features on a component. However, these are ambiguous in the sense that the indications given thus far do not define exactly where the tolerances apply. In traditional 2D views, especially with line tolerances that apply to edges, it is naturally implied that tolerances will be in the plane of the view. However, when tolerances apply to surfaces or surface elements, there is no indication of how the tolerances should be validated. Sometimes it is useful, therefore, to specify exactly where the tolerances apply and hence where they should be measured and validated. To do this we require the use of datums, which will be covered in Chapter 6, and we also need some further symbolic means of specifying the locations of the measurement profiles with regard to the datums. These are presented in Chapter 7.

The increasing use of 3D views on drawings and of using **product manufacturing information** has also highlighted the need to bring 3D and 2D methods into alignment, and this was one of the major motivations behind the ISO 1101: 2017 standard.

6 Datums

So far, the coverage of dimensions and tolerances has related to constraints on feature size and form. To fully constrain geometric features, however, we must turn our attention to positional tolerances, those that relate to a feature's orientation and location. To facilitate this, it is convenient to use **datums**, features to which the position of other features can be readily established. This chapter is dedicated to a basic description of datum features as these are fundamental to the operation of the ISO GPS scheme. **ISO 5459:2024** gives the current detail of how datums are to be defined and utilized. This standard was under review for some time, and superseded the 2011 version recently. A second standard, **ISO 2692:2021** gives details on how datums should be used with material condition modifiers such as MMR, LMR and RPR.

HISTORICAL NOTE

Position tolerancing methods have probably been in use for some time, for example, as implied limits that were part of company policies, and some people have reported seeing explicit tolerances stated on drawings from around the early 1900s; however, the author can find no evidence of this. What we do know is that John V. Liggett, in his 1970 book *Fundamentals of Position Tolerance* [1], reported that the first-known formal method of tolerancing was developed at the Royal Torpedo Factory in Alexandria, Scotland, by Stanley Parker. There is good reason to believe this claim because Liggett was an expert in geometric tolerancing and a contemporary of Parker's. One of Parker's friends and mentors, F. H. Rolt, who spent time working in the Scottish shipping and shipbuilding industries at the P. & O. Steam Navigation Company at Caird of Greenock and at the Fairfield Shipbuilding & Engineering Company in Govan, Glasgow, explained that Parker's methods of dimensioning and tolerancing were used to re-dimension drawings for the production of a naval weapon (presumably a torpedo). In 1940, Stanley Parker's report to the British Admiralty, on formal positional tolerancing, was published under the title *Notes on Design and Inspection of Mass Production Engineering Work* [2] and in 1956 he published the book *Drawings and Dimensions* [3].

6.1 DATUMS AND DATUM SYSTEMS

Some further definitions are important in any discussion about datums.

DOI: 10.1201/9781003395072-6

6.1.1 Datum

A datum is a **theoretically exact reference**. They are used to establish the **orientation** and **location** of other features. Datums need not exist physically but may be given a physical model by reference to a **real** feature on a part or **material holding device**, for example. Datums may be found on **nominal**, **real** or **derived** features.

6.1.2 Datum System

A datum system is a collection of situation features that are jointly used to establish an environment for validating features.

6.1.3 Datum Feature

A datum feature is a real (physical) feature on a part that can be used to establish a model of a datum with an **associated** feature.

In a previous chapter we defined the concept of an associated feature as being an ideal feature that is fitted to the measured or scanned data of a real feature using one of a number of possible algorithms (e.g. minimax, least squares, maximum inscribed or minimum circumscribed). These features are used to establish a datum. If using scanned data from **coordinate measuring machines (CMMs)**, then the size, position and orientation of associated features can readily be calculated.

6.1.4 Datum Simulator

A datum simulator is a physical object that can be used as a model for a datum, and therefore it can be measured from. Examples include **flat** or **surface plates,** or **tables, angle plates**, **gauge pins** or **chucks** for example, and these are commonly made from cast iron or granite. They are often necessary when using traditional measuring gear such as **height gauges** or **profile gauges**. In effect, physical simulators are a form of associated feature, but with a much more limited repertoire of features with which to model, for example, **least-squares fits**, are not possible.

6.1.5 Situation Feature

A **situation** feature is a point, line, surface or combination of these that is used as a reference for the position and orientation of a feature; for example, a sphere has a situation feature that is its centre point. These features are the geometric entities that are used to define a datum.

Figure 6.1 shows the situation features for a range of basic feature types which are in turn based on what ISO calls **invariance classes** (**plane, cylinder, helix, sphere, revolute, prismatic and complex**). Table 6.1 outlines the nature of the situation feature for each invariance class.

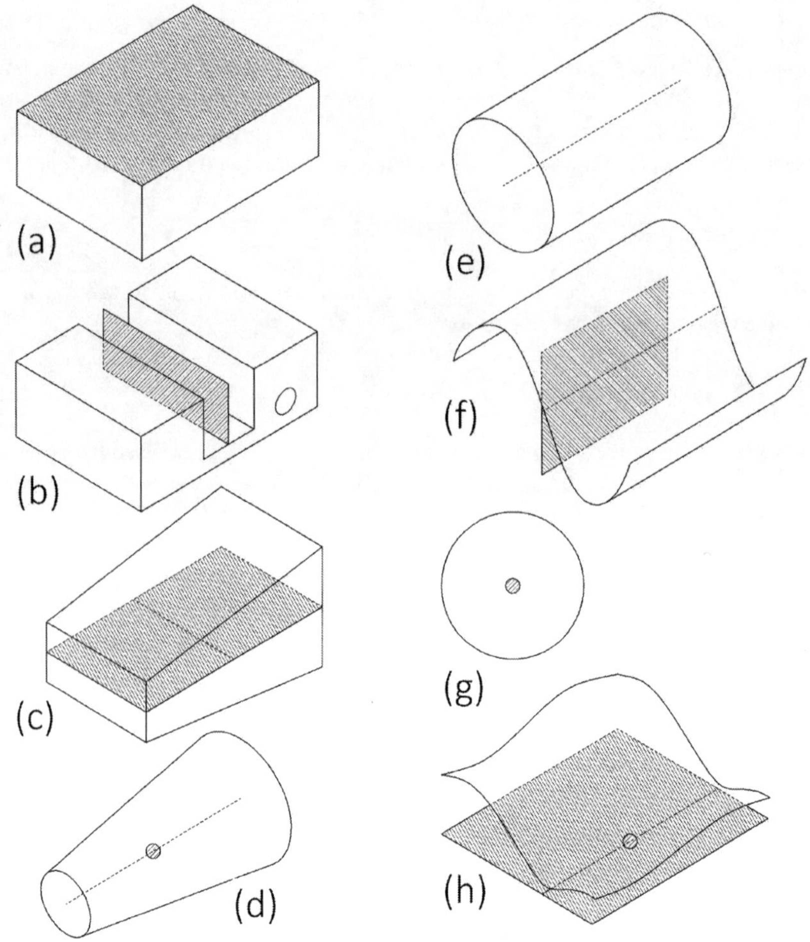

FIGURE 6.1 Situation features for invariance class features.

TABLE 6.1
Invariance Class Features and Their Situation Feature Types

Invariance Class	Situation Feature(s)
Plane (a), (b)	Plane
Prismatic (c)	Plane and line
Revolute (d)	Line and point
Cylinder (e) (can be used for helical surfaces)	Plane and line
Extruded profile (f)	Line
Sphere (g)	Point
Complex (h)	Plane, line and sphere

However, we identify features on a component to act as a datum and we can simulate datums with certain types of inspection equipment that we can measure from, for example, surface plates or gauge pins. Part datums can be defined on features, parts of features and across multiple features, and they can be combined into complex datum systems. To give an overview of geometrical tolerancing methods, we will use some relatively simple examples for now and deepen our coverage later.

6.1.6 Display of Datums

Datums can be depicted on 2D and 3D PMI visualizations. There are a number of ways they can be indicated, as shown in Figure 6.2.

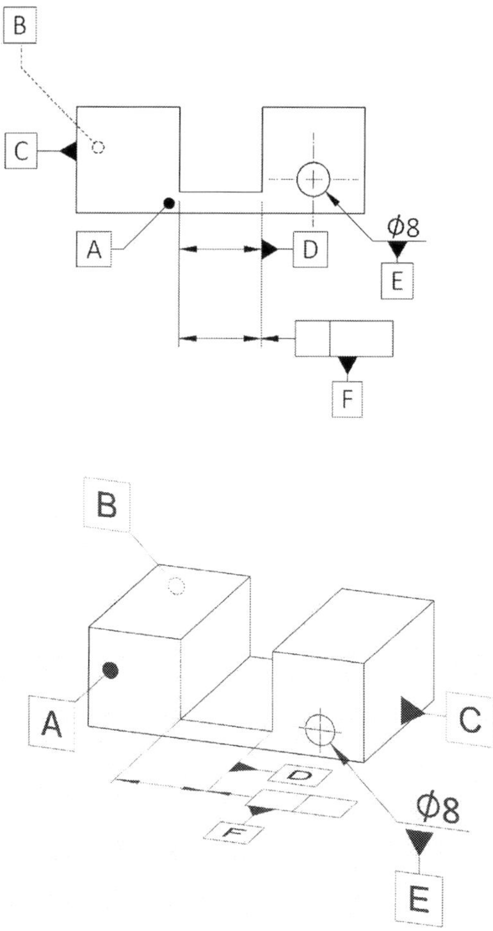

FIGURE 6.2 Display of common datum symbols.

6.2 DATUM SYSTEMS

As the basic idea of a datum is to enable the location and orientation of features so that their measurements can be specified and validated, we should consider how datums can also be combined into **datum systems** that help constrain a feature's variance so that it is locked in position, whether virtually or physically.

6.2.1 Degrees of Freedom

Figure 6.3 shows the commonly used nomenclature for the six **degrees of freedom (DoFs)** based on the norms for many CAD systems. There are three translational (X, Y, Z) and three rotational (Rx, Ry, Rz).

We can establish planar and rotational datums on features on a drawing that will act as reference features, from which we measure and hence show the required dimensions of a part. Subsequently when the part is manufactured we can place the physical part on various fixtures and establish physical surfaces from which physical measurements can be taken (simulated datums) to validate the manufacture. If parts are scanned with laser scanners or point clouds established from CMM data, we can also use software to establish associated datums.

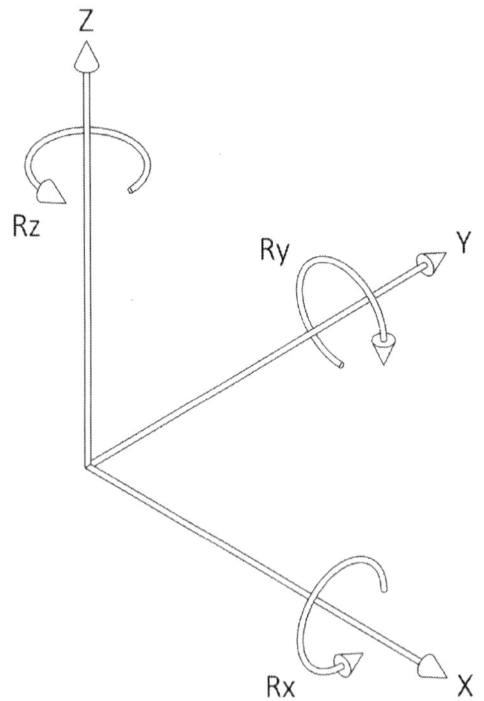

FIGURE 6.3 Six degrees of freedom (DoFs).

6.2.1.1 Example 1—Prismatic Part

Using devices such as surface tables and angle plates, we can consider an example by constraining a prismatic part on planes. Figure 6.4 shows a design of a simple part that has three datum planes defined on it. In ISO terminology the computer-modelled part is known as the nominal part.

We see that a large flat face with a flatness tolerance of 0,01 mm is chosen as datum A. The long-side face must be perpendicular to datum A within a tolerance zone of width 0,01 mm and is chosen as datum B. Lastly, the face shown at the bottom of Figure 6.4 is chosen as datum C, and it must be perpendicular to both A and B, also within a tolerance zone of 0,01 mm wide.

Suppose the part is manufactured and is to be validated to make sure it is within tolerance. In ISO terminology such a physical part is said to be a **real part,** and it has real features. Figure 6.5 shows a model of a possible real part where the face deviations have been exaggerated.

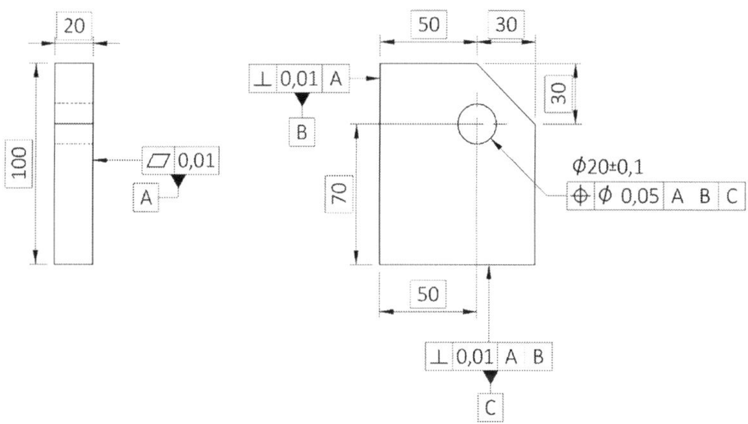

FIGURE 6.4 Nominal part with datums.

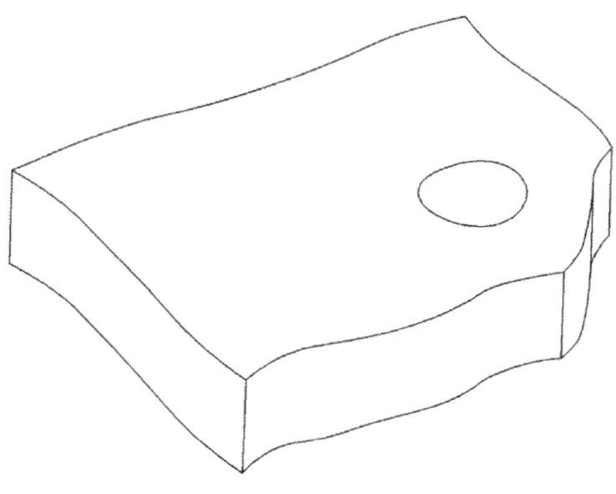

FIGURE 6.5 Real part with exaggerated deviations.

To measure the dimensions of the real part we can imagine constraining it using a surface table and two angle plates. Figure 6.6 shows the real part shortly before such an operation. It is shown 'floating' because it is considered to have six DoFs (X, Y, Z, Rx, Ry, Rz) and thus has free movement of location and orientation.

Assuming the bottom face (datum feature A) is of an acceptable shape we can place the real part on the surface table as shown in Figure 6.7.

Once the real part was placed on the surface table with the face associated with datum A down, it immediately lost three DoFs. It is on the table so it can no longer be moved in the Z direction, and it cannot rotate around the X or Y axes, so it has lost Z, Rx and Ry DoFs. Next, we might slide the part across the plate until it is up against the angle plate as shown in Figure 6.8.

Now that the part is considered flat against the angle plate, it has lost a further two DoFs. It can no longer move in the Y direction and cannot rotate around the Z axis so it has lost Y and Rz DoFs.

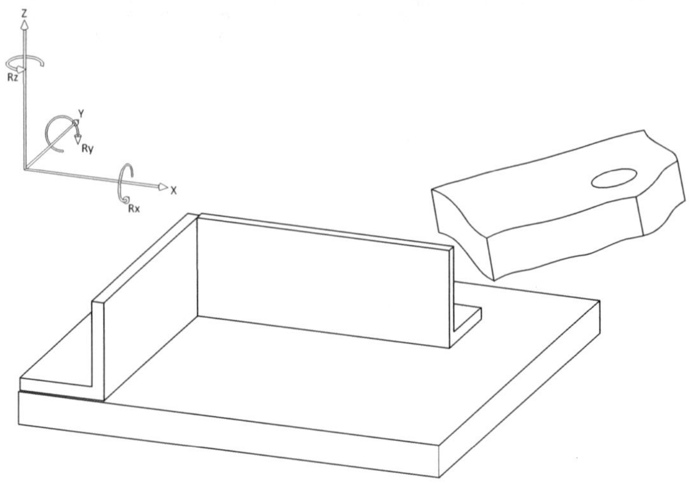

FIGURE 6.6 Real part about to be constrained for measuring.

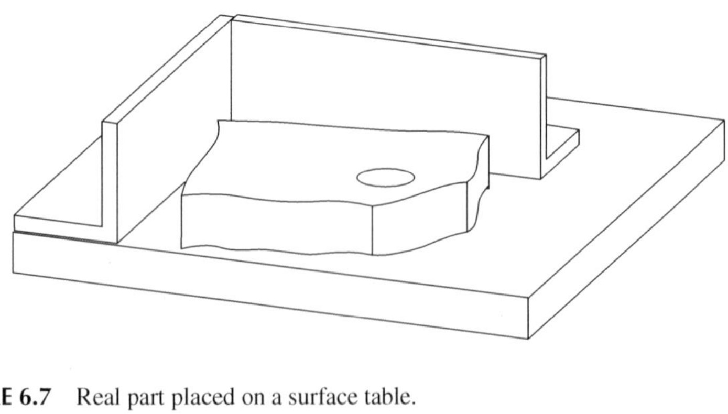

FIGURE 6.7 Real part placed on a surface table.

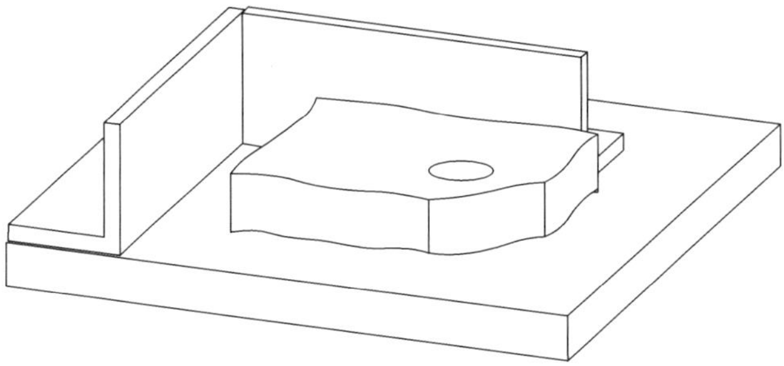

FIGURE 6.8 Real part on table aligned to an angle plate.

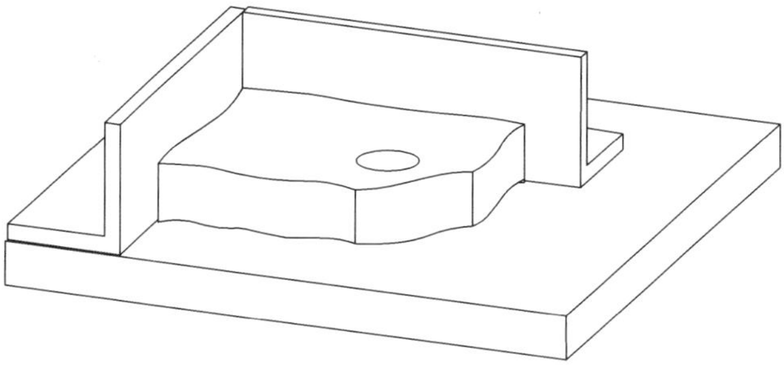

FIGURE 6.9 Real part fully constrained.

Lastly, let us assume we push the part in the X direction until it touches the second-angle plate (which is perpendicular to the first) as shown in Figure 6.9. The part has now lost the X DoF.

Now the part is both fully oriented and located and we consider the surface table and angle plates as simulators for the datums and take measurements from them to features on the real part. We might do this with metrology equipment such as callipers, depth gauges, dial (profile) gauges, measurement arms or CMMs, for example.

6.2.1.2 Example 2—Rotational Part

Figure 6.10 shows a simple rotational part with its nominal and real forms.

To constrain such a part we might use vee blocks as well as surface plates and angle plates. Figure 6.11 shows the rotational example in such a set-up.

This set-up uses vee blocks to constrain translation and rotation in relation to the Z and Y axes and so immediately removes four DoFs (Z, Y, Rz, Ry). By touching the angle plate the part also has translation in the X direction constrained. This still allows rotation around the X axis (Rx) and this is sometimes acceptable but the addition of clamping devices will commonly restrict this too. In this example the angle plate will act as a datum simulator for datum A. If rotation has to be under better control the part could also be placed in a chuck.

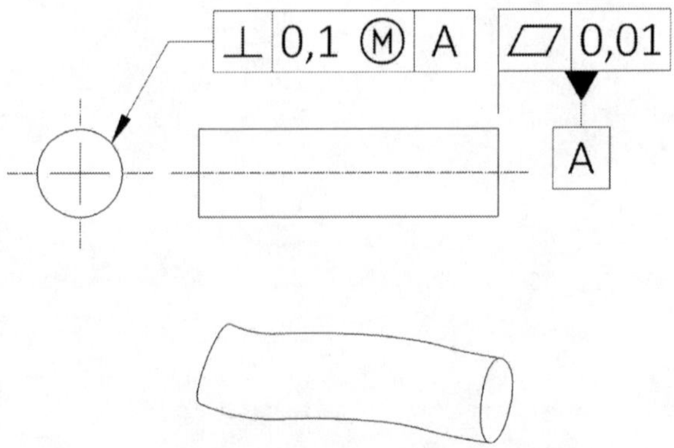

FIGURE 6.10 Nominal and real views of the rotational part.

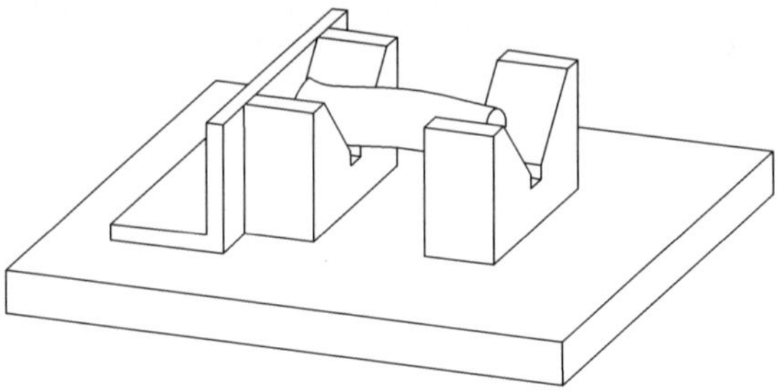

FIGURE 6.11 Real rotational part constrained for measuring.

6.3 PROBLEMS WITH COMMON VALIDATION METHODS

In both of the previous examples, clamping devices may need to be added to hold the parts in their constrained positions so that physical measurements might be taken without dislodging the parts.

There are however significant further practical problems in trying to constrain and validate parts in the manner described earlier.

6.3.1 How Flat Is a Surface (Flat) Plate?

In the previous examples we assumed that the surface plate was perfectly flat and that the angle plates were perfectly flat and perpendicular to the surface plate. In many practical situations this is a valid assumption in the sense that any deviations in the plates will be very small in comparison to those on the part being measured. What makes an acceptably flat surface to serve as a surface plate is not always entirely clear but should have a grade specified and agreed between a part supplier and customer. This can be done according to national or other standards for conformance and calibration, for example, BS 817 (UK) or DIN 876 (Germany).

Typically a surface table of 400 × 250 mm might be expected to have a permitted **deviation from flatness** of 4μm **for Grade 0** (high accuracy), 8μm **for Grade 1** (general inspection), 16μm **for Grade 2** (marking out) and 32μm **for Grade 3** (general support plate and low grade marking out) work. Marking out is the process of marking a workpiece with, for example, scribed lines to indicate cutting areas.

> **Reminder**
>
> Datum—a theoretical point, line or surface with perfect form shown on a part model/drawing.
>
> Datum feature—an actual derived or integral feature on a real part that has been assigned a datum
>
> Associated datum—perfect form feature constructed from measured data on a datum feature
>
> Datum simulator—a physical device that will act as a datum and can be measured from, for example, a surface table

We stated earlier that the previous location and orientation operations could be carried out easily only if the shape of the features would allow it. Consider a flat plate like the one shown in Figure 6.12. If the part is placed on a surface table with the surface feature A down and it has a generally concave form, it will be stable and the datum may be assumed to lie on the 'nearest contacting line or surface', as shown.

If, however, the part is largely convex, such as that shown in Figure 6.13, it may rock and have several configurations with which it can contact the plate.

Situations like this can cause challenges in validation. Options to deal with such problems might involve the improved specification of the datum surface that is contacting the datum simulator or compensating for the deviations using multiple set-ups to characterize the datum surface. **Shims** can be used to support the part where necessary. If the part form, or shape, is likely to lead to difficulties then a note on the drawing or PMI can be used to state any assumptions or to advise on strategies for convex surfaces.

If scanning equipment is used that can conveniently provide point data around the full region of interest then it may be possible to create associated datum planes in software.

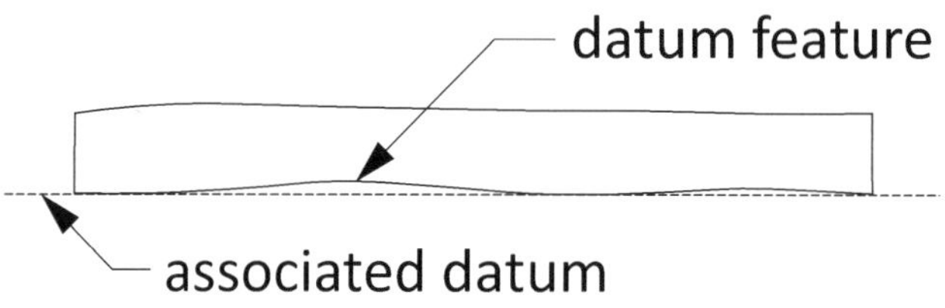

FIGURE 6.12 Plate with concave datum feature.

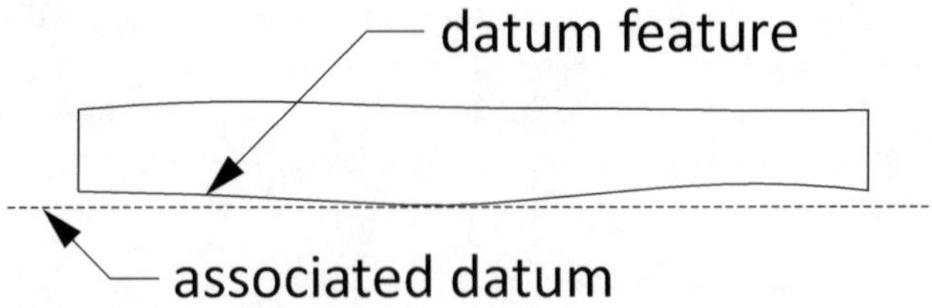

FIGURE 6.13 Plate with convex datum feature.

6.4 COMBINED AND CONSTRAINED DATUMS

So far, the datums we have considered have related to entire single features, whether integral or derived; however, it can be convenient to combine features to form a datum or to constrain the datum status to a specific region.

6.5 COMMON AND COMBINED ZONES

Sometimes it is useful to define a datum in terms of more than one feature. One way to do this is with a common datum as shown in Figure 6.14. The two datums share a common axis and when the real part is made and measured, the two resulting associated features are defined with perfect alignment (co-linear axes) although they may be of different diameters. The real part might be held between two chucks to set up such a measurement arrangement and relationships to other features assessed for conformance to any stated geometric tolerances, for example, **run-out**, which we cover in more depth in a later chapter.

Combined Zones can also be defined on parts to create datums. Figure 6.15 shows an arrangement where three flat surfaces together define a datum plane. In this case the three features can be regarded as one, for example, as if there are no gaps between them. If the 'CZ' symbol is omitted the three features would each have to meet a 0,2 mm tolerance zone independently. When this is the case the letters 'SZ' should be used to avoid any ambiguity.

6.6 CONSTRAINED ZONES: DATUM TARGETS

Sometimes it is not possible, or at least not convenient, to create a datum that will span an entire surface or set of surfaces due to uncertainty about the exact positions of the places we would want to measure from. For example, parts that have been cast or moulded may have surfaces that are substantially different from their modelled depictions. Sheet metal parts and other relatively flexible parts might also be better measured from precise positions rather than referring to whole surfaces. In such cases we can use **datum targets**. These are precise locations on a drawing that we can use to position datums on the measurement devices being used. Datum targets can be assigned to points, lines or surface regions. Figure 6.16 shows common depictions of datum targets.

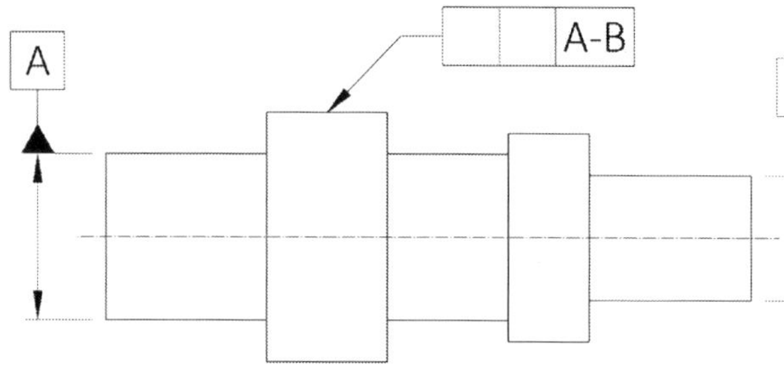

FIGURE 6.14 Combined rotational zones.

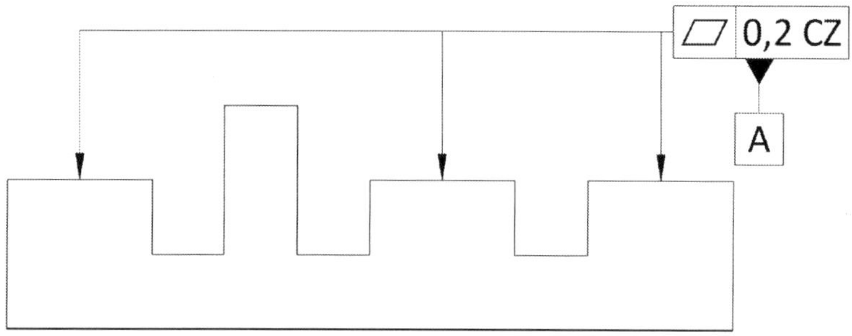

FIGURE 6.15 Combined zones from multiple features.

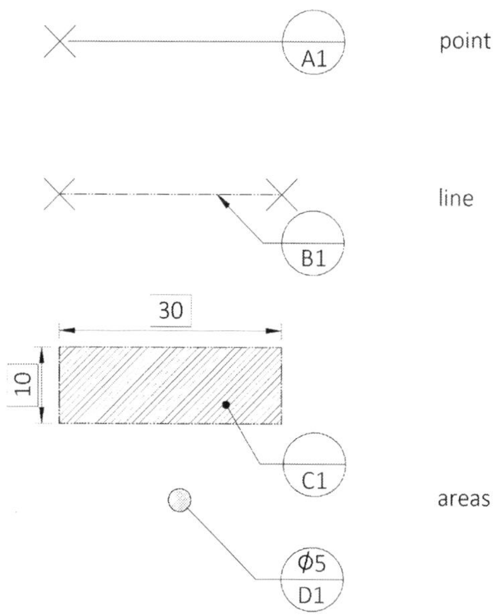

FIGURE 6.16 Common depictions of datum targets.

6.6.1 Example—Datum Targets on a Flat Plate

Figure 6.17 shows a triangular plate with datum targets depicted on the plan view. The three targets combine to define the datum A, and this is shown in the side elevation. Note that the datum identifier is followed by the details of the individual targets that it includes. Note also that the positions of the datum targets are defined by theoretically exact dimensions (TEDs).

6.7 CONTACTING FEATURES

It is not always possible, or desirable, to establish datums from associated features on a part. Sometimes the part may have to be set up for validation, and possibly finally assembled onto other parts which do not share the same geometric shape. An example would be where a spherical part might be inserted into a cone as shown in Figure 6.18.

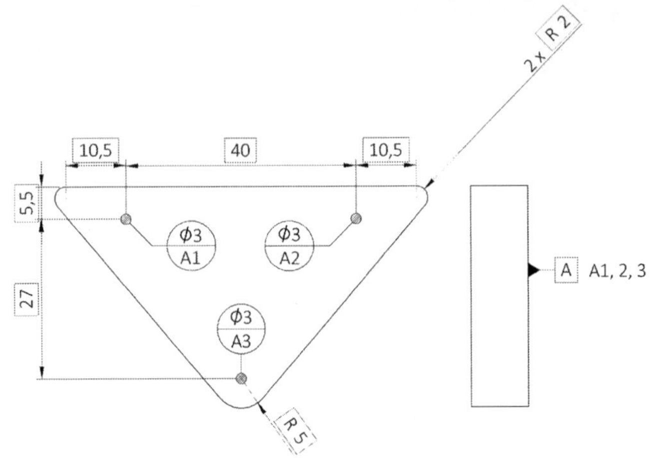

FIGURE 6.17 Datum targets on a planar face.

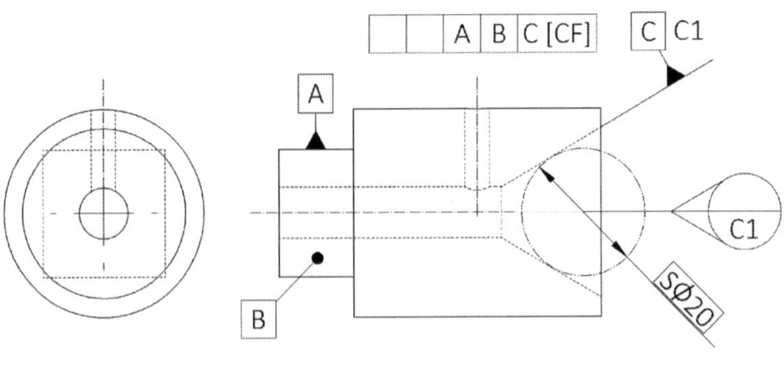

FIGURE 6.18 Datum system with moveable datum target.

In this case a contacting feature (sphere) is used to establish the datum C. The exact location of C is, in the example, unknown, and so a moveable datum target is used to locate the centre of the sphere and hence the exact location of datum C at the time of validation. Dimensions can be placed referencing the sphere centre.

6.8 THEORETICALLY EXACT DIMENSIONS (TEDS)

The reader may have noticed dimensions shown inside rectangular boxes on drawings or within this book already. These are **TEDs** and are, as the name implies, exact and with no tolerances attached directly to them. They form an important part of the ISO GPS philosophy. Distances or angles between features, often between feature and datum feature, are shown as TEDs, while any tolerances are in the form of geometric tolerances, which can be attached directly to the features being controlled.

6.8.1 Examples: Traditional and GPS Style Positional Tolerancing

Figure 6.19 shows a typical method for how a hole feature might be located using traditional ± tolerances (which are very large for illustration purposes rather than any realistic application). Note that a fuller explanation of **location** tolerances will be given in Chapter 8.

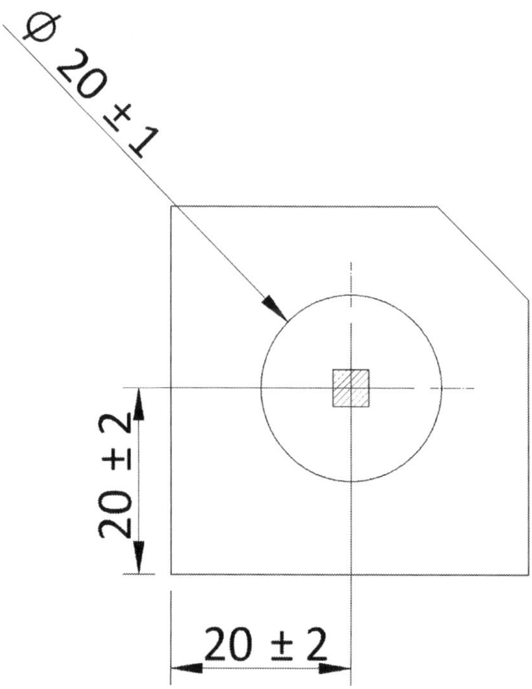

FIGURE 6.19 Traditional ± positioning of a hole feature.

The tolerancing system shown implies no means of manufacture of the part or how it should be validated once made. If, however, we imagine the common situation where such a part might be made by locating a drill bit at the position shown with a typical milling/drilling/boring machine, then the drill bit centre point can locate anywhere within the square shaded 4 × 4 mm region shown. Then if the drill bit size and drilling operation successfully meets the diametrical tolerance allowances of ±0,1 mm then the part will be acceptable.

Figure 6.20 shows a very similar part that has the hole feature located by TEDs but with a geometrical locational tolerance on the feature.

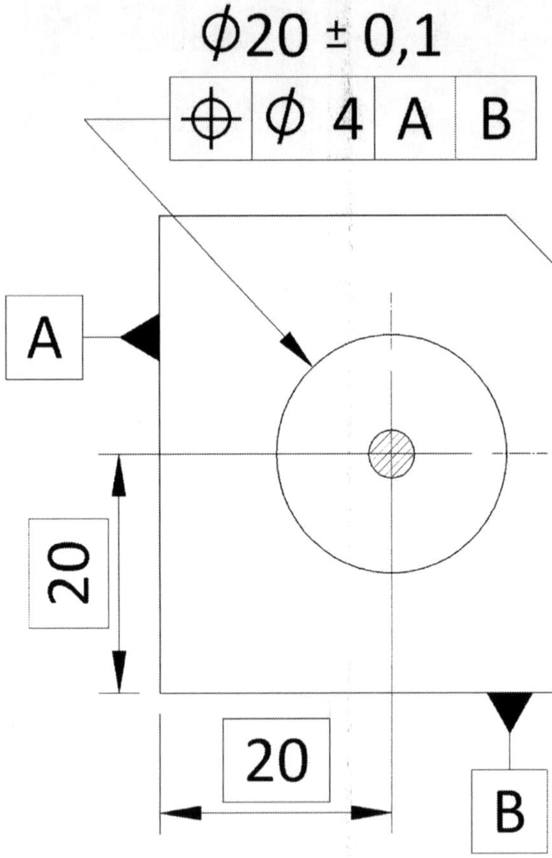

FIGURE 6.20 TEDs and GPS positioning of a hole feature.

Note that this example is incomplete. According to the ISO GPS rules when the system is invoked on any part of a drawing it must be used throughout, so for completeness we should apply some geometric tolerances to the datum features and complete the dimensioning and tolerancing of the other surface features. Using only a 2D depiction it is also implied that the hole central axis should lie perfectly perpendicular to the part's base.

The example shown in Figure 6.19 produces a similar result to that shown in Figure 6.18 except that the tolerance region for the positioning of the hole centre is now circular rather than square. If we consider the maximum radial deviation allowable, it is the same for both examples (the hypotenuse of the 4 × 4 tolerance zone = 5,66 mm). Sometimes it is claimed that this gives a 57% increase in tolerance area using a positional tolerance with a diameter symbol over a conventional ± tolerance area, for the same allowable radial deviation. However, whether this can be realized practically depends on the nature of the application and the manufacturing methods used.

6.9 REFERENCES

[1] Liggett, John V.; "Fundamentals of Position Tolerance", New York, NY: Society of Manufacturing Engineers, 1970.
[2] Parker, Stanley; "Notes on Design and Inspection of Mass Production Engineering Work", Sheffield: Naval Ordnance Gauge Factory, 1940.
[3] Parker, Stanley; "Dimensions and Tolerances", Pitman, 1956.

7 Orientation Tolerances

7.1 INTRODUCTION

To fully constrain a feature on a part, it is necessary to apply limits on the orientation (and location) of the feature. Orientation constraints in the ISO system come in the form of tolerances of **parallelism**, **perpendicularity** and **angularity**, and again we can also use **line profiles** and **surface profiles** as defined in **ISO1101:2017** and shown in Table 7.1.

Profile tolerances, as we have seen, can be applied without a datum, but in such cases they can only control the form, or shape, of the feature and not how it is oriented or located. By placing a dimensional reference to a datum, a profile tolerance can then be used to control **form** and **orientation** (and **location**—see Chapter 8).

7.2 INDICATORS ADJACENT TO THE TOLERANCE INDICATORS

In previous versions of geometric tolerancing systems, there were sometimes ambiguities in how a tolerance might be validated on the finished component. For example, the 2D depiction of the profile tolerance shown in Figure 7.1 does not specify where exactly the measurements of the elements of the profile should be made, although it is implied that the profile lines would be parallel to the profile as it is shown. The lower 3D depiction in Figure 7.1 makes this explicit by adding an **indicator adjacent** to the **tolerance indicator** that shows that any measurements should all be in planes intersecting the part and parallel to the datum A.

Note that the profile tolerance in the example only controls the form (or shape) of the united features between L1 and L2 but does not explicitly say where the profile should lie exactly. In other words, the whole tolerance zone defined by the profile between L1 and L2 is free to float as a rigid body as long as it is oriented parallel to face A. We will shortly show how profiles can be constrained in space by their orientation and in the following chapter by their location.

Note also that the UF indicator is not strictly necessary because the required geometry is implied by the fact that the two surfaces (flat and curved) are tangentially connected.

Additional indicators adjacent to the tolerance indicators can make design intent clearer by specifying where validation measurements should be made and so ensure the feature meets its functional requirements. The additional indicators consist of further symbols and previous references or adjacent to the tolerance indicator and will identify where any validation measurements should be taken. These extensions are themselves references to datum features (except **ACS**) that may be revolute, cylindrical or planar. A cylinder is clearly also a revolute feature, but they are distinct in the ISO 1101 standard. The five additional indicators are shown in Table 7.2.

DOI: 10.1201/9781003395072-7

TABLE 7.1
Symbols for Orientation Tolerances

Symbol	Name
//	Parallelism
⊥	Perpendicularity
∠	Angularity
⌒	Line profile
⌓	Surface profile

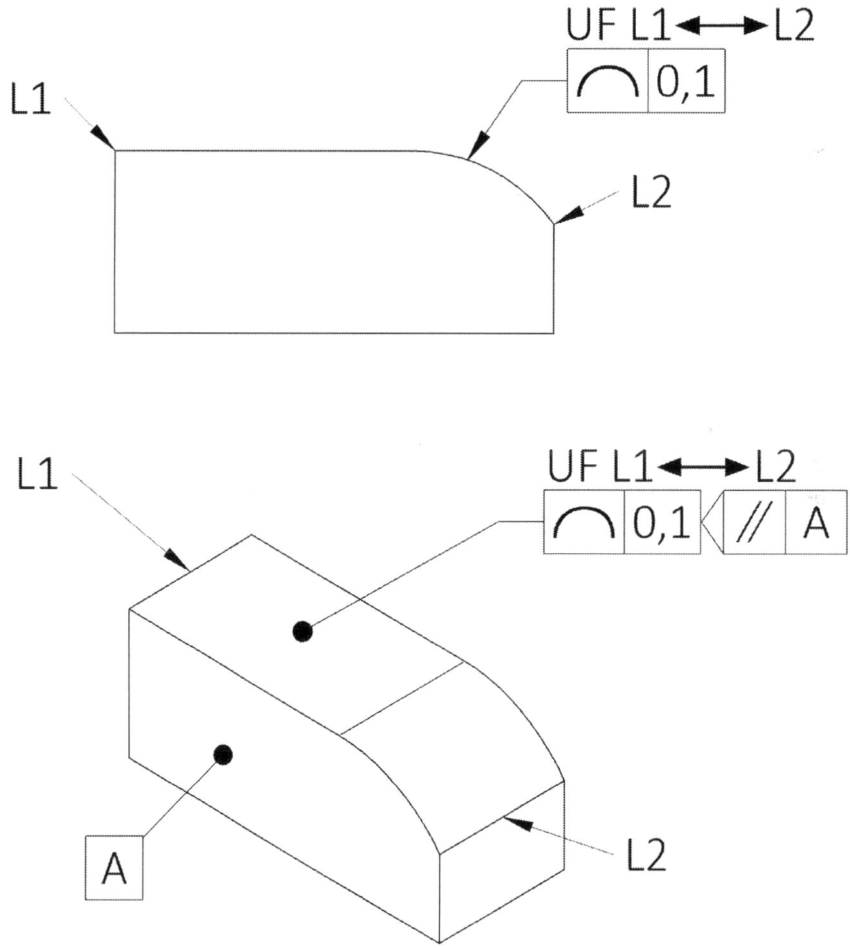

FIGURE 7.1 Additional indicator adjacent to the tolerance indicator.

TABLE 7.2
Indicators Adjacent to the Tolerance Indicator

Extension	Name	Symbols Available	Description
ACS	Any cross-section	Placed above the tolerance frame	The tolerance should be met at any cross-section measured, only necessary if there is ambiguity.
	Intersection plane	∥ ⊥ ∠ ≡	Identifies planes for line and profile-based validation measurements.
	Orientation plane	∥ ⊥ ∠	Identifies a plane that defines the orientation of a **tolerance zone (TZ)**, for example, where a planar TZ is positioned on a median line or a cylindrical TZ on a point.
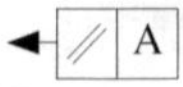	Direction feature	∥ ⊥ ∠ ↗	Identifies the direction of the width of a tolerance zone; it is normal to the feature profile by default.
	Collection plane	↗ ⊥ ∠	Identifies a plane passing through a collection of surfaces used for validation measurements, only of use with **all-around** indicators.

7.3 PARALLELISM TOLERANCE INDICATOR

Simple parallelism tolerance zones consist of two parallel planes that contain the toleranced feature and these are themselves parallel to the referenced datum feature. Any measurements made to validate the toleranced feature should be made according to explicit directions given in indicators adjacent to the tolerance indicators. It is often convenient to combine two parallel tolerance zones that are normal to each other to create an emergent square or rectangular sectioned zone. It is also common to make the axis of cylindrical features parallel to each other and to modify the tolerance zone by preceding the tolerance value with a diameter symbol to signify that the tolerance zone should be round or cylindrical in form.

Parallel tolerances only relate to linear features, that is, straight lines and flat surfaces, but these may be either integral (physical surface) or derived (centre line or plane) features. Consider some example cases.

7.3.1 Example—Plane Parallel to Planar Datum with Independence Principle

Figure 7.2 shows a simple rectangular part where a specification is made for parallelism from a face to a datum while respecting the principle of independence.

7.3.2 Example—Plane Parallel to Planar Datum with Envelope Principle Invoked

Figure 7.3 shows a simple rectangular part where a specification is made for parallelism from a face to a datum while invoking the envelope principle.

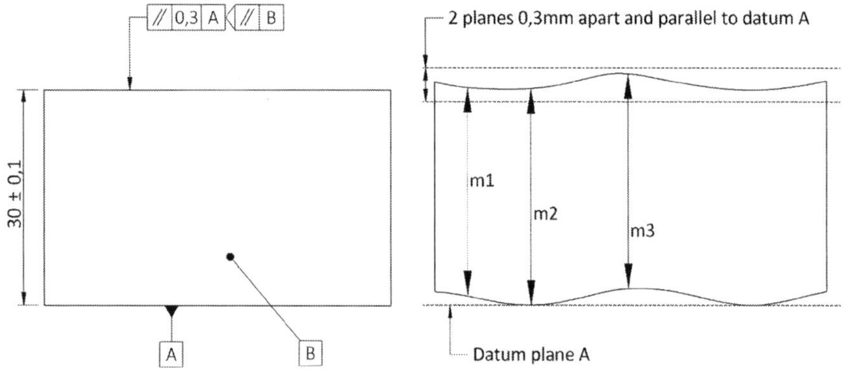

FIGURE 7.2 Plane parallel to planar datum plane.

Notes:

- The **intersection plane** indicator adjacent to the tolerance indicator shows that all measurements of parallelism to datum A should be made on profiles that lie on planes that are parallel to datum B.
- The principle of independence is the default in ISO standards and means that each set of two-point measurements (m1, m2, m3 . . .) must be within the size tolerance range of 29,9–30,1 mm. However, the parallel tolerance can be any value and therefore greater than the 0,2 mm width of the size tolerance.

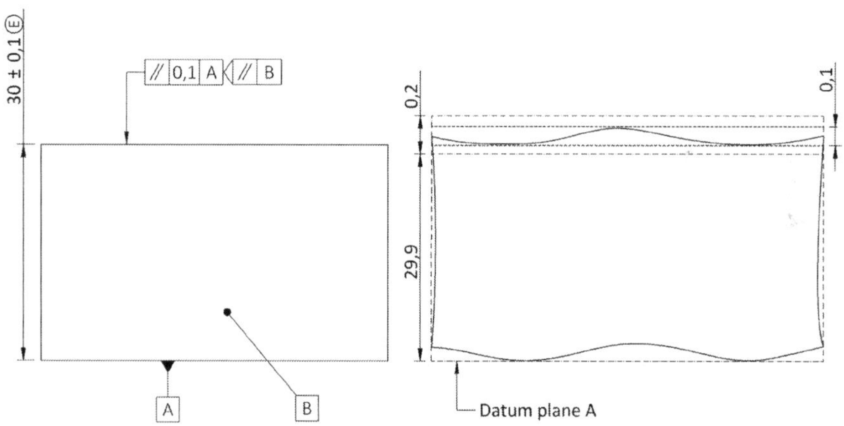

FIGURE 7.3 Plane parallel to planar datum plane with envelope principle invoked (see size tolerance).

Notes:

- The intersection indicator adjacent to the tolerance indicator shows that all measurements of parallelism to datum A should be made in planes that are parallel to datum B.
- The envelope symbol added to the size tolerance invokes the envelope principle. This means that not only must all size measurements be within the size tolerance range of 29,9–30,1 mm but the whole part must also be able to fit into a slot that is 30,1 mm wide. This in turn means that the parallel tolerance zone must be within the size tolerance zone and therefore cannot exceed the size tolerance.
- The placing of the datum plane chosen in the example is based on a **tangent plane** and is the datum's **situation feature**.

7.3.3 Example—Centreline Perpendicular to Plane

Figure 7.4 shows two holes on a part where the larger hole's centreline should be parallel to the smaller hole's centreline (datum A) within a tolerance zone that consists of two planes 0,02 mm apart and which are perpendicular to datum B.

7.4 PERPENDICULARITY TOLERANCE INDICATOR

Simple perpendicularity tolerance zones commonly consist of two parallel planes that contain the toleranced feature and these are themselves perpendicular to the referenced datum feature. Like parallel tolerances, any measurements made to validate the toleranced feature should be made according to explicit directions given in indicators adjacent to the tolerance indicators. Also, like parallel tolerances, two tolerances can be used to form rectangular sectioned zones and diameter symbols used to create cylindrical zones.

Perpendicular tolerances only relate to linear features, that is, straight lines and flat surfaces, but these may be either integral (physical surface) or derived (centre line or plane) features. Some examples follow.

7.4.1 Example—Centreline Perpendicular to Datum Plane A Parallel to Datum B

Figure 7.5 shows a block with a protrusion toleranced with a perpendicularity indicator.

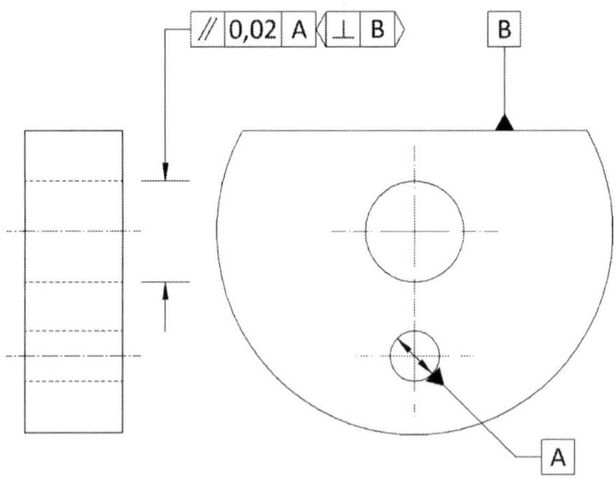

FIGURE 7.4 Centreline (derived feature) tolerance zone perpendicular to plane (integral feature).

Notes:

- The orientation indicator adjacent to the tolerance indicator shows that all measurements of parallelism to datum A should be made in planes that are perpendicular to datum B.

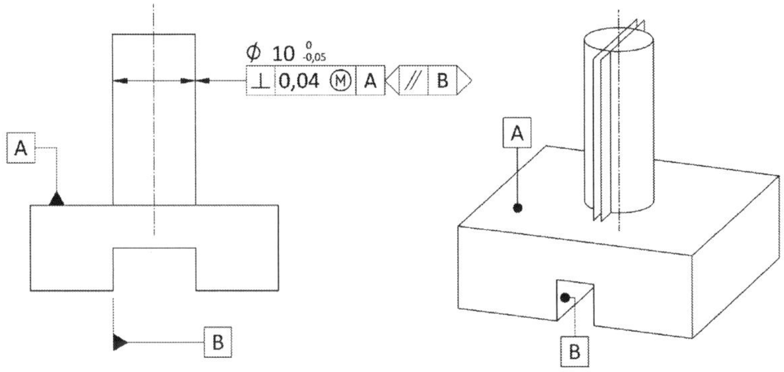

FIGURE 7.5 Centreline perpendicular to datum plane A and between two planes parallel to a second datum plane B.

Notes:

- The illustration on the left shows a specification that includes a size tolerance of +0,0/–0,05 mm and a requirement that the shaft centreline should be perpendicular to datum A in any plane parallel to datum B within a tolerance zone that is 0,04 mm wide.
- The illustration on the right shows a 3D interpretation of the specification. The specification only requires perpendicularity between the tolerance zone, which is parallel to datum B. We could add a second specification with an adjacent indicator that requires a further tolerance zone perpendicular to datum B, and this would mean a rectangular sectioned tolerance zone would be required.

7.4.2 Example—Centreline Perpendicular to Datum Plane A

In the example shown in Figure 7.6, a diameter symbol is used to indicate that a cylindrical tolerance zone for the protrusion centreline is required, which is perpendicular to datum plane A.

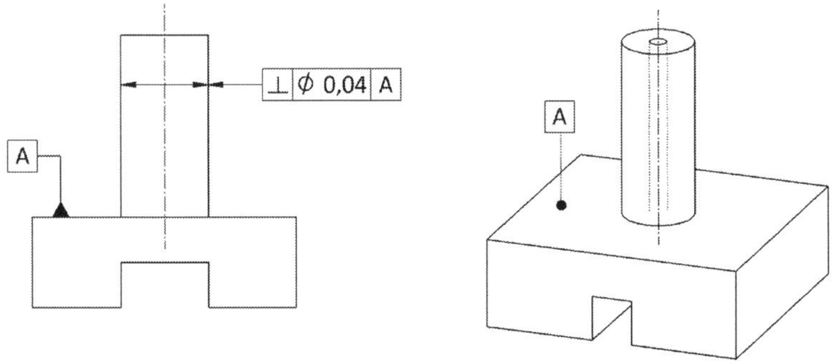

FIGURE 7.6 Centreline in a cylindrical tolerance zone perpendicular to plane.

Note:

- By adding a diameter symbol in the tolerance indicator (left illustration), the specification requires that the tolerance zone should be cylindrical as shown in the illustration to the right.

7.5 ANGULARITY TOLERANCE INDICATOR

Angularity tolerances are similar to both parallel and perpendicularity tolerances but represent a more general case. They typically consist of two parallel planes that contain the toleranced feature, and these are at an angle to a referenced datum feature. Like parallel and perpendicular tolerances, any measurements made to validate the toleranced feature should be made according to explicit directions given in indicators adjacent to the tolerance indicators. Again, like parallel and perpendicularity tolerances, two tolerances can be used to form rectangular sectioned zones and diameter symbols used to create cylindrical zones.

Angularity tolerances also only relate to linear features such as straight lines and flat surfaces, and these may be either integral (physical surfaces) or derived (centre line or plane) features. Some examples follow.

7.5.1 Example—Hole at an Angle to a Plane

Figure 7.7 shows an example of hole that is to be made at an angle to a base plane (datum A) and parallel to a side plane (datum B). Typically, using conventional workshop metrology equipment such a part might be validated by first checking the flatness tolerance on datum A and then placing the part with datum A contacting a surface (flat) table. The part would then be slid along to an angle plate set at 90° such that datum B would be flat against it. The required validation measurements could then be made from the datum simulators (the table/plate). In many cases a third datum surface could be specified to fully locate the part prior to validation, for example, on an end face.

Figure 7.8 shows how the previous example might be depicted using a 3D view.

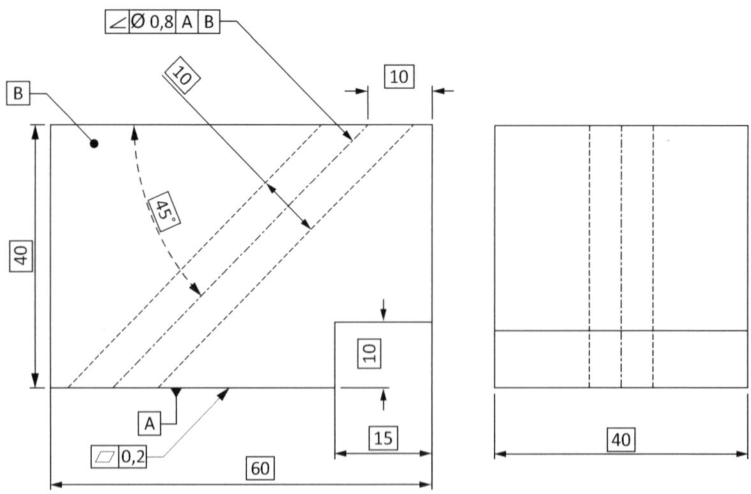

FIGURE 7.7 Hole at an angle to a plane in 2D.

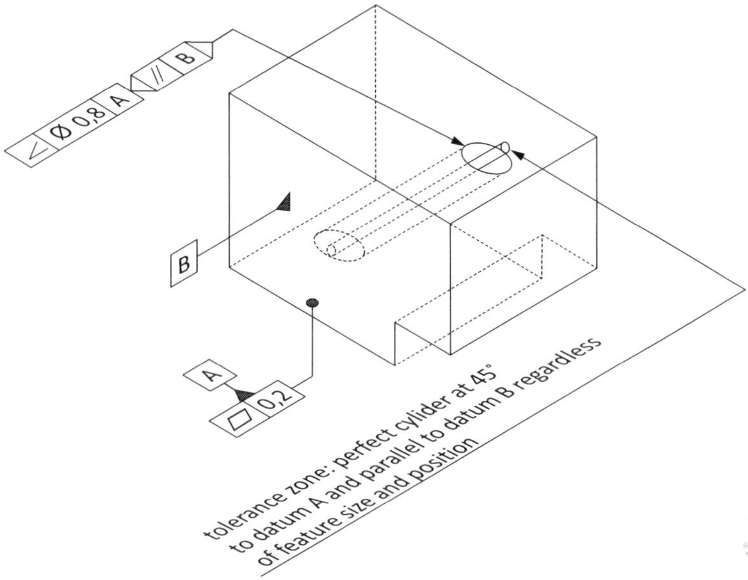

FIGURE 7.8 Interpretation of angularity tolerance in Figure 7.7.

Note:

- The diameter symbol in the tolerance indicator precedes the value 0,8 mm, making the tolerance zone a cylinder.

7.5.2 Example—Integral Feature (Plane) Angled to a Datum Plane

Figures 7.9 and 7.10 show a simple example of a part with a face angled and tolerance to a datum.

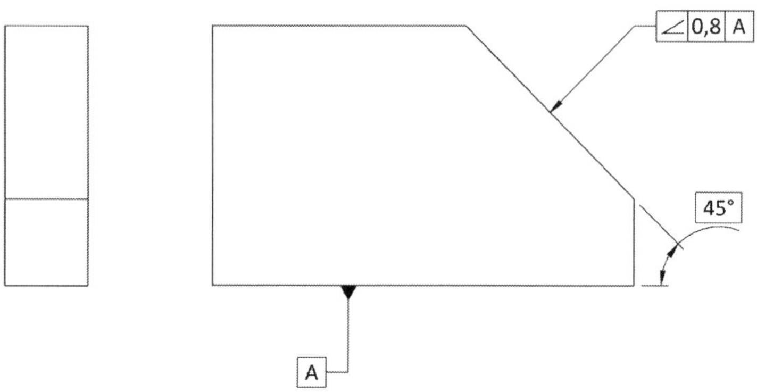

FIGURE 7.9 Plane surface angled to datum plane.

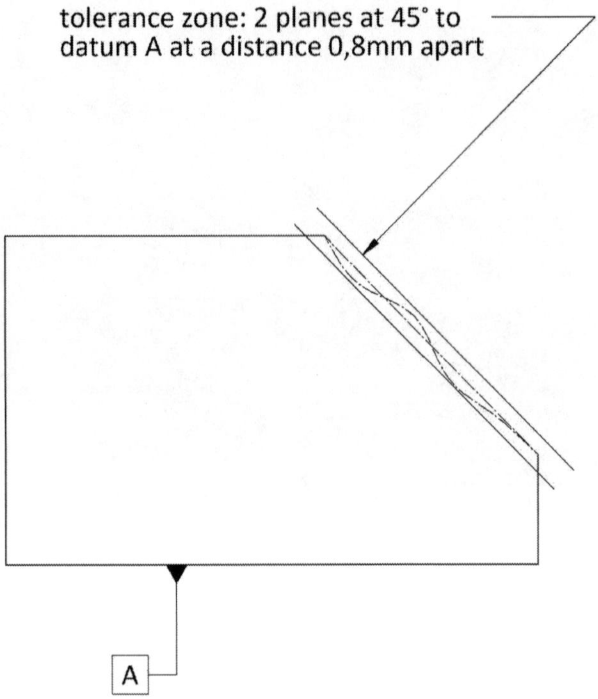

FIGURE 7.10 Interpretation of plane angularity tolerance in Figure 7.9.

Note:

- The drawings show no TEDs, leaving the position of the angled plane undefined.

7.6 LINE PROFILE TOLERANCE INDICATOR

In Chapter 5 we introduced line profile and surface profile tolerances constraining only the form, or shape, of a profile line or surface. We can however add orientation data or location data to a line or surface by providing references to a datum or datums. Line profile tolerances are 2D in form. They can be applied to surfaces, but they only then represent single line profiles in a given direction upon that surface.

7.6.1 Example—Profiles with Form Tolerances and with Form and Orientation Tolerances

Figure 7.11 shows a profile of a line with its form constrained and additionally with its orientation and locational constrained.

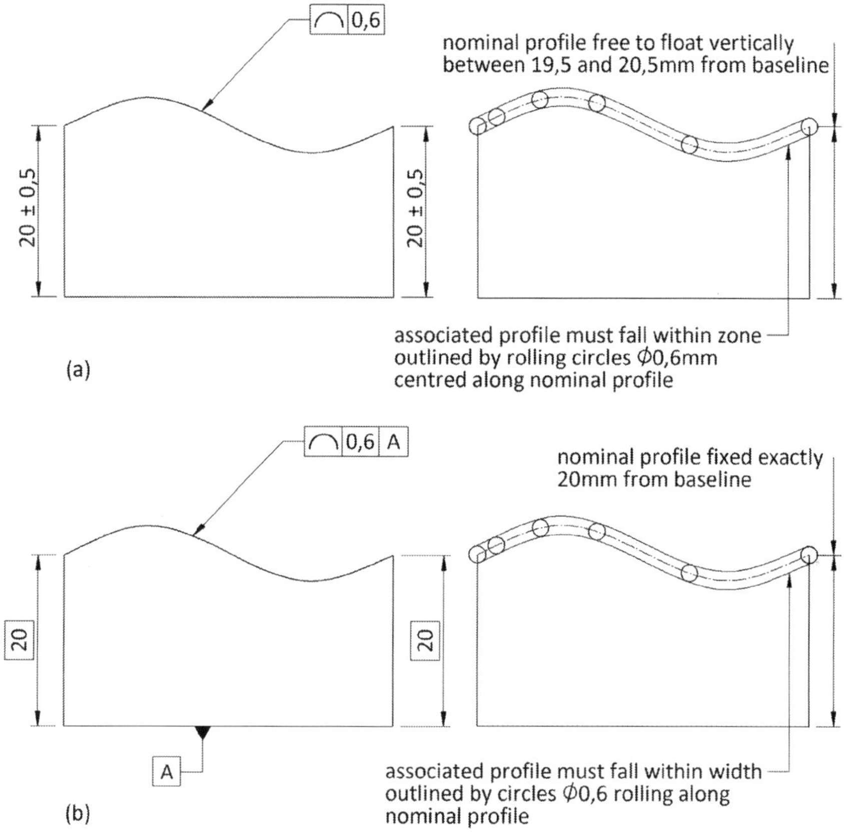

FIGURE 7.11 Profiles constrained by form and orientation.

Notes:

- In Figure 7.11(a) the profile has only a form geometric tolerance. This leaves the nominal profile free to float within the ± tolerance zone giving some control over orientation and location.
- In Figure 7.11(b) the profile has been constrained exactly by the TEDs and therefore has its orientation (and location) locked.
- The drawing does not give exact information about the form of the profiled feature. This might be a series of tangency continuous radii or it may be defined by interpolation and/or approximation curves. **ISO 129–1** demonstrates the use of ordinate dimensions (see Chapter 2) for curves as one method of description. However, it is possible to fully define complex curves using control points and an algorithm with which to calculate any point on the curve (or surface). At present there is no ISO standard for one single method of doing this, but a discussion of some of the possibilities is given in Chapter 10.

7.6.2 Example—Profile Tolerance Indicator with Adjacent Indicator

The example shown in Figure 7.12 shows a line profile tolerance applied to a profile.

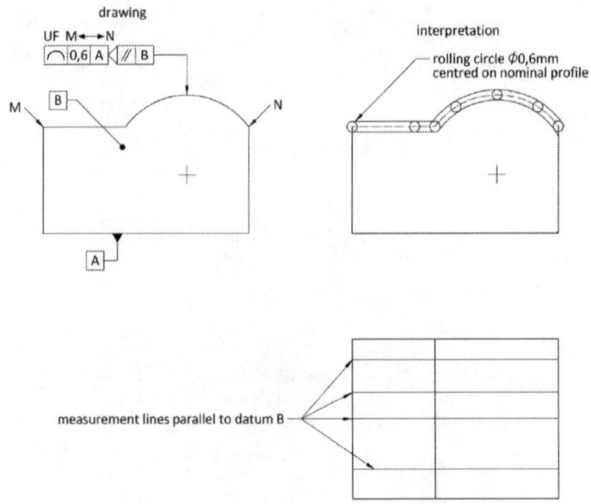

FIGURE 7.12 Line profile tolerance with adjacent indicator.

Notes:

- The tolerance indicator shows a profile tolerance of 0,6 mm, which corresponds on the view on the right as a rolling circle profile with diameter 0,6 mm and centre on the nominal profile.
- The letter callouts for M and N show two points on the part profile. The UF designation shown above the tolerance indicator means that two or more features are to be considered as one for tolerancing. The M between N symbol above the tolerance indicator shows the positions between which the individual features should be measured for the line profile tolerance shown on the drawing.
- On a 2D view it would be normal to assume that any measurements of lines on the top surface would be taken normal to the drawing view in which the tolerance appears. The addition of the adjacent indicator showing that all profile measurements should be taken parallel to datum B makes this explicit and removes any ambiguity about orientation of the part. This is particularly useful on 3D PMI views and is recommended in **ISO 1101** as a means of bringing 2D and 3D methods into alignment.

7.7 SURFACE PROFILE TOLERANCE INDICATOR

Surface profiles work similarly to line profiles in that they are 3D tolerances and apply to entire surfaces or parts thereof. Thus instead of using a rolling circle indicator, **ISO1101** indicates the use of a rolling sphere that follows an entire surface (i.e. infinite number of profiles along and across a surface).

Like a line profile tolerance, a surface profile is a very powerful tool that can be used to constrain parts in a great number of ways, often instead of using other tolerances such as parallelism. Also, like line profile tolerances, surface profile tolerances can be used to constrain form, orientation and location.

Let's consider some examples.

7.7.1 Example—Surface Profile of a Unified Feature

Figure 7.13 shows two segments that form a unified feature. The tolerance zone can be created by rolling an imaginary cylinder centred on the nominal profile of the surface.

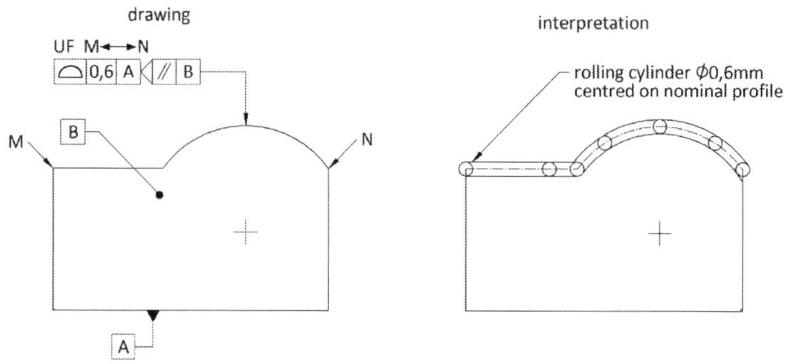

FIGURE 7.13 Surface profile of a unified feature.

Note:

- ISO 1101 states that the tolerance zone is limited by two equidistant surfaces that envelope an infinite set of spheres that are centred on the nominal surface, and in this case this is equivalent to sweeping a cylinder centred on the nominal profile, the tolerance zone being created by the boundary of the solid that is swept.

7.7.2 Example—Surface Profile of a United Feature with 'All Around' Symbol

Figure 7.14 shows a second example of a part with a united feature indicator above the tolerance indicator. In this case the 'all around' indicator, a circle, has been placed

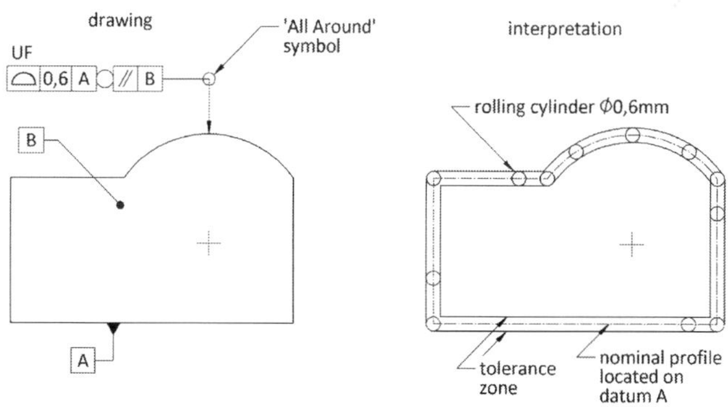

FIGURE 7.14 Surface profile of a united feature with 'all around' symbol.

Notes:

- The orientation indicator shows that any profile to be measured should be parallel to datum B. This clarifies the orientation of the 'all around' specification.
- The resulting tolerance zone is formed by rolling a cylinder of diameter 0,6 mm around the prescribed profile with its centre always on the nominal profile. This results in a tolerance zone that has rounded corners.

at the junction of the tolerance zone leader line and break line. This means that the tolerance zone goes completely around the outline of the part in the view shown.

7.7.3 Example—Surface Profile with a Combined (Tolerance) Zone

A commonly used alternative to using a united feature is to use what is termed a 'combined zone'. In this case the tolerance zones are imagined for the individual features and then combined into a single zone as shown in Figure 7.15.

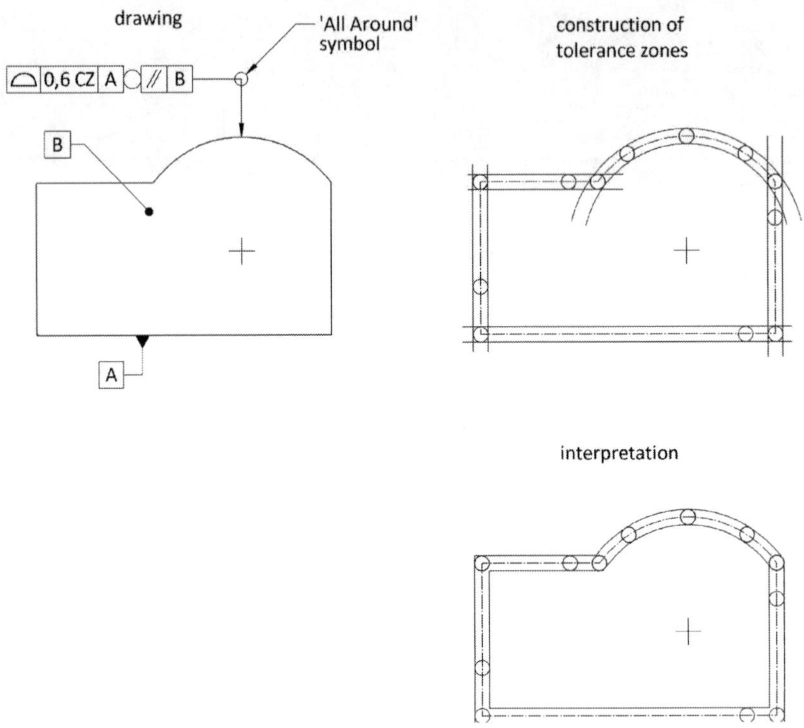

FIGURE 7.15 Surface profile with a combined zone.

Notes:

- Individual tolerance zones are created and extrapolated from the features and then subsequently combined. ISO 1101 does not specify extrapolation methods, but these are straightforward for line and radii if not for complex (e.g. B-spline) surfaces.
- The ISO standard specifies that there should not be discontinuities where the individual zones are merged.
- The resulting tolerance zones have sharp edges at their corners (unlike the UF example in Figure 7.14).

7.7.4 Example—Surface Profile with Unequal Tolerance Zone

So far, all the examples have assumed that the tolerance zones have been equally distributed around the nominal line or surface; however, these can be unequally distributed. In assemblies it is common to model two parts (say, a bolt and threaded hole) at their nominal sizes but also to arrange the tolerances so that the bolt has only a negative tolerance while the hole has only a positive tolerance. This may be necessary to ensure a clearance on the assembly.

Figure 7.16 shows how an unequal zone can be specified in the ISO GPS system.

Comments

The material presented in this chapter gives an introduction to the use of orientation symbols used in feature control frames to constrain the orientation of tolerance zones. Traditional style ± tolerances also give some control over the location of tolerance zones and hence of the features. In the following chapter, specific location tolerance symbols will be used in feature control frames to fully constrain features and their tolerance zones.

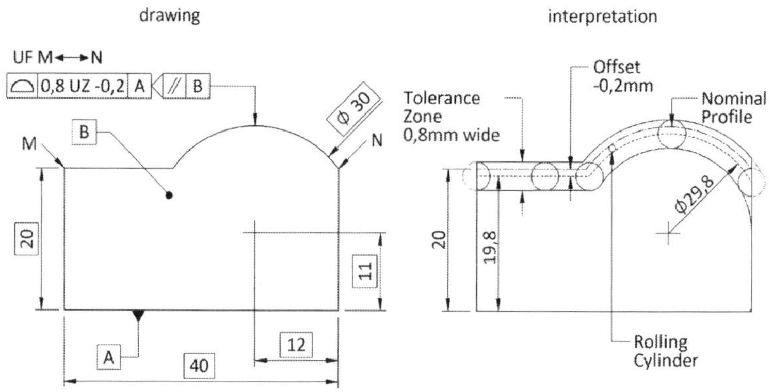

FIGURE 7.16 Surface profile with unequal zone.

Notes:

- In the tolerance indicator the UZ symbol is placed beside the tolerance width value to indicate that the zone is to be unequally distributed around the nominal surface profile.
- The value of –0,2 mm represents the amount by which the tolerance will be offset and the negative value specifies that it will act to reduce the material size.
- The tolerance zone is generated by the space of an infinitely rolling sphere of diameter 0,2 mm following the contour of the **theoretically exact shape (TEF)**, in this case described by the nominal profile in the 2D view. Because it is a simple 2D surface profile we can also represent the action of the sphere as a rolling cylinder. This produces a profile known as the tolerance zone median line, and it differs in shape from the nominal profile since the cylinder produces rounded corners on the outside of bends.
- The tolerance zone is equally distributed about the tolerance zone median profile.

8 Location Tolerances

8.1 INTRODUCTION

Location tolerances are probably the most intuitive tolerances after simple feature of size tolerances. They represent the location of key features in space and thus involve referencing these to datum features. The features used for location can relate to functional features on a part, or they can be features that are added to the part purely to aid location and orientation. Such features are commonly employed in automated systems to help locate and orient parts for manufacturing operations such as stamping or assembly as well as gauging and other validation activities (not to forget packaging). Location tolerances can be applied to a wide variety of feature types, both integral (edges, surfaces) and derived (centre points, centre lines, median surfaces).

Location tolerances include the ubiquitous **position tolerance**, which we will study shortly, as well as **concentricity, coaxiality, symmetry** and **line/surface profile** tolerances (see Table 8.1). As we saw in Chapter 7, profile tolerances can be used in a number of ways and can constrain features that several other tolerance types (e.g. parallelism, perpendicularity) also cover. The combination of position and profile tolerances can cover a very wide range of applications, and some engineers claim to use these for the vast majority of their work. Let's consider basic positional tolerancing.

8.2 POSITION TOLERANCE INDICATOR

Position tolerances can refer to integral or derived features, including points and centre points, lines and centre lines, surfaces, profiles and median profiles. Lines can be straight or nonstraight, and surfaces can be flat or non-flat. **ISO 1101** states that the shape of the toleranced feature must be explicitly described on either a drawing or a reference to a CAD model geometry. Let's consider some examples.

TABLE 8.1
Symbols for Location Tolerances

Symbol	Name
⌖	Position
◎	Concentricity
◎	Coaxiality
⌯	Symmetry
⌒	Profile any line
⌓	Profile any surface

DOI: 10.1201/9781003395072-8

8.2.1 Example—Spherical Feature

Figure 8.1 shows how a position tolerance symbol can be used in a **feature control frame** (sometimes referred to as a tolerance indicator) to locate the position of a spherical feature.

8.2.2 Example—Hole Position

Figure 8.2 shows a simple method of locating a hole feature using a position tolerance.

8.2.3 Example—Angular Position

Figure 8.3 shows how a cylindrical feature with a planar end face set at a given angle might be located.

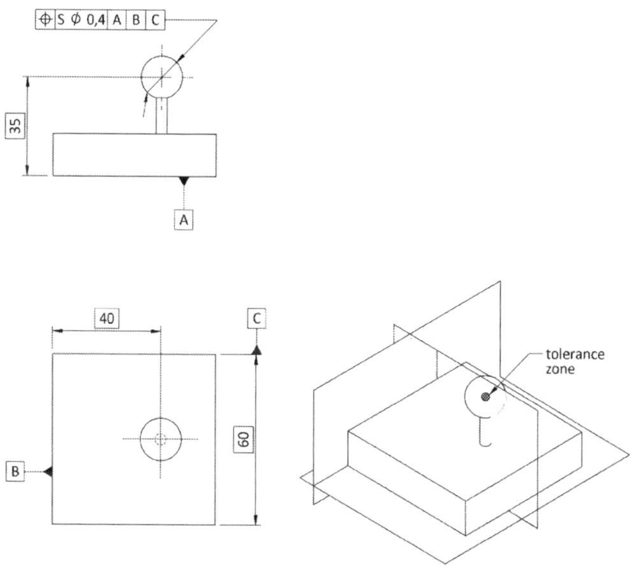

FIGURE 8.1 Spherical tolerance zone.

Notes:

- The 'S' symbol before the diameter symbol indicates that the tolerance zone is a sphere of diameter 0,4 mm.
- The centre of the tolerance zone is located exactly 35 mm from the planar datum A, 40 mm from the planar datum B and lying on the planar datum C.
- Datum C is defined as being midway between two surfaces on the part.
- The result is a specification that, when the associated feature of the sphere is considered, its centre must lie within the tolerance zone described. Remember that associated features are those created from validation measurements made on the physical part as described in Chapter 5.

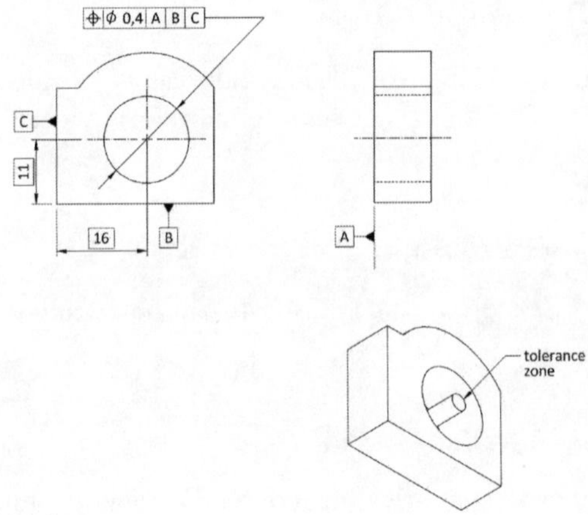

FIGURE 8.2 Hole positioning

Notes:

- The diameter symbol in the feature control frame indicates that the tolerance zone will be a cylinder 0,4 mm wide.
- The cylindrical tolerance zone's axis will be perfectly perpendicular to datum A and perfectly parallel to datums B and C at distances of 11 and 16 mm, respectively.

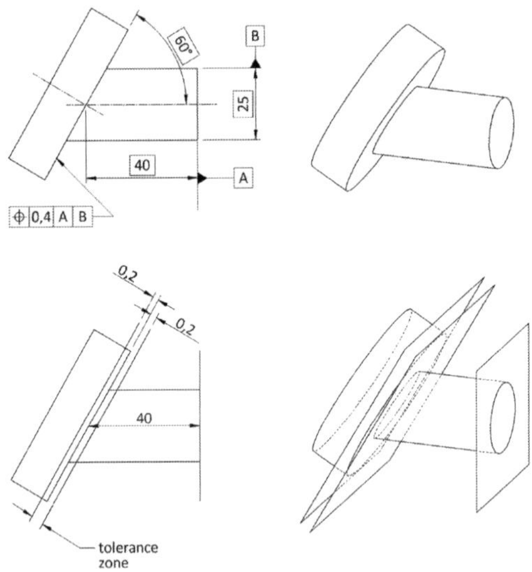

FIGURE 8.3 Positioning of angular feature

Notes:

- The positional tolerance represents two planar faces angled at exactly 60° from the axial datum B and at a distance 0,4 mm apart.
- The tolerance zone planes are symmetrically offset from the nominal feature's face whose centre point is exactly 40 mm from the planar datum A.

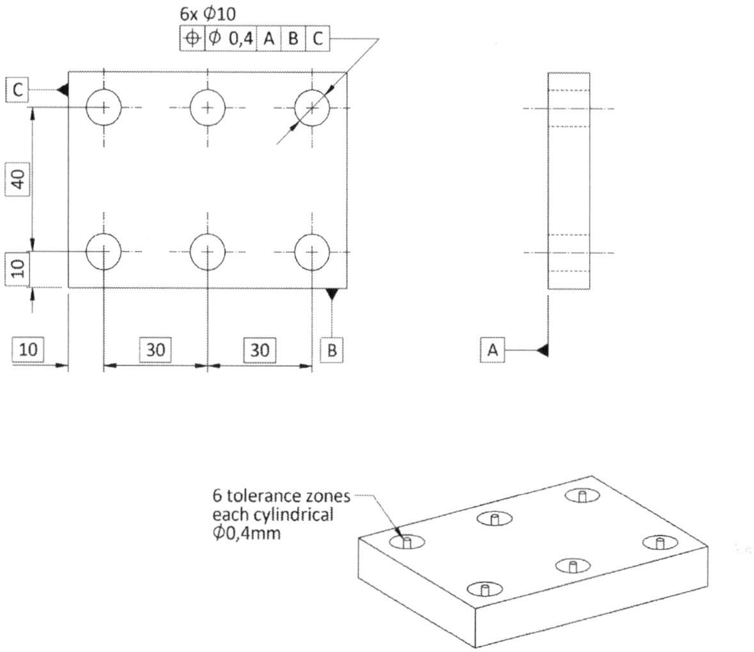

FIGURE 8.4 Position of patterned holes.

Notes:

- The holes are independent but each constrained by the single tolerance frame.
- Each hole's central axis must fall within a tolerance zone that is cylindrical, 0,4 mm wide and perpendicular to datum A as well as parallel to datums B and C at the distances given by the TEDs on the drawing.

8.2.4 Example—Pattern

Figure 8.4 shows how the features in a pattern can have their locations set with a single feature control frame.

8.3 CONCENTRICITY TOLERANCE INDICATOR

Concentricity tolerances relate to circular elements and are therefore 2D in nature. They are particularly useful in constraining pipes where there may be some deviation axially along the pipe but where the wall thickness needs to be controlled. An example will demonstrate how this can typically be achieved.

8.3.1 Example—Concentricity

Figure 8.5 shows a short section of pipe that might be subject to pressure, or used in shielding electrical conductors, for example.

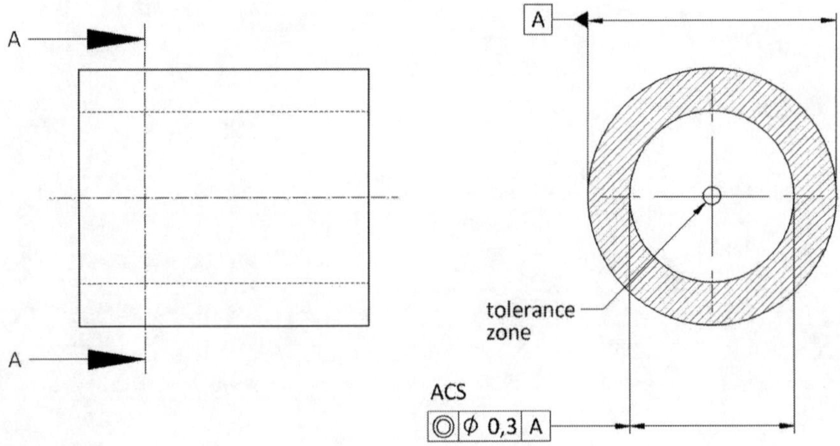

FIGURE 8.5 Concentricity of circular cross-sections.

Notes:

- The diameter symbol indicates that the tolerance zone will be circular. It is 0,3 mm wide and centred on the ideal axis defined by datum A.
- The 'ACS' symbol indicates that the tolerance is specified for any cross-sectional part of the pipe. Each such measurement is independent of any others.

8.4 COAXIALITY TOLERANCE INDICATOR

Coaxial tolerances relate to cylindrical elements and are therefore 3D in nature. They are particularly useful in applications where alignment of cylindrical elements is important, for example, with turned components where moments of rotation need controlled. An example will show how coaxiality tolerances can be specified.

8.4.1 Example—Coaxiality

Figure 8.6 shows a part with two cylindrical surfaces that must be commonly centred along their axes.

8.5 SYMMETRY TOLERANCE INDICATOR

Perhaps the least commonly used of the location indicators is the symmetry tolerance indicator. It is useful in specifying the symmetric location of features such as slots; however, the same effective tolerance zones can be specified using position tolerances, for example. As a result, it is absent from the latest ASME standards. An example of the use of symmetry tolerances is given in the following subsection.

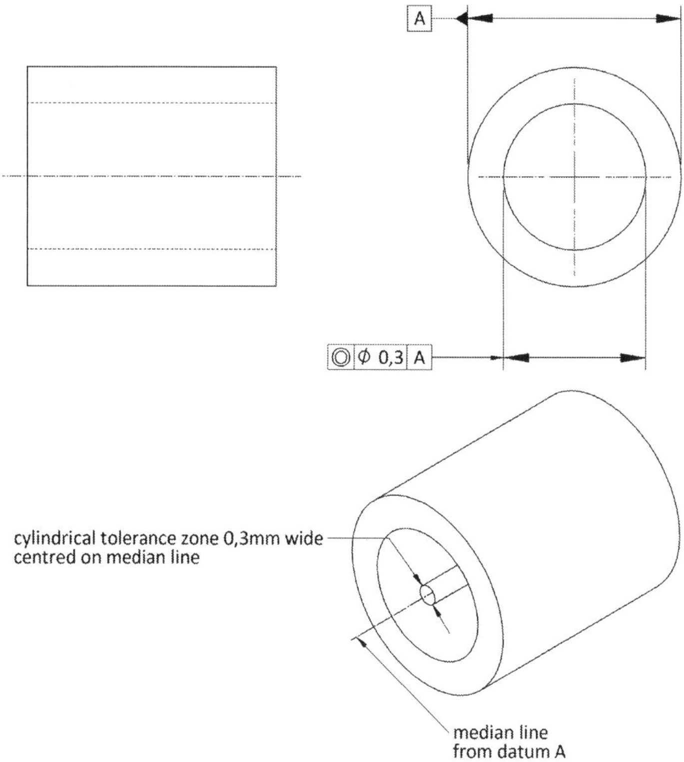

FIGURE 8.6 Coaxial cylindrical features.

Notes:

- The diameter symbol indicates that the tolerance zone will be cylindrical and 0,3 mm in diameter. Its axis is set on the median line of the datum feature A.
- The result is that any the associated feature of the inner cylindrical surface on the part must have its median line within the tolerance zone.

8.5.1 Example—Symmetry of a Slot

Figure 8.7 shows how a symmetry tolerance can be used to position a slot feature.

8.6 COMMENTS

Location tolerances almost complete the geometrical tolerances that are in common use for general components. In the next chapter we will investigate the use of run-out tolerances which are used mainly for turned parts. Before we finish with general methods, it is worth reminding ourselves what the examples given in this chapter actually mean.

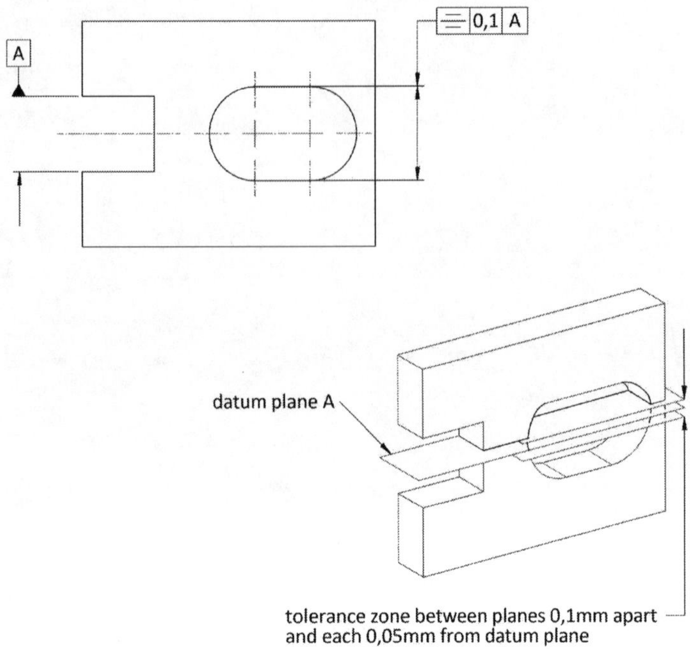

FIGURE 8.7 Symmetric tolerance zone.

Notes:

- The tolerance zone consists of two parallel planes that are 0,1 mm apart and symmetrically distanced from a median plane defined by datum A.

The TEDs show the theoretic locations and sizes of ideal nominal features while the feature control frames describe any allowable variations in form, orientation or location with reference to a datum feature or datum system. The datum planes shown in the interpretation of examples show the planes that validation measurements might be taken from. These may, in practice, be generated in a number of ways. For example, a planar face might in reality be a datum simulator such as a flat table surface on which the part is lying. Alternatively, if the part has been scanned or sampled, it may be a surface fitted using least-squares or Chebyshev methods as described earlier in the book.

9 Run-Out Tolerances

9.1 INTRODUCTION

Run-out tolerances are composite tolerances since they can constrain several attributes of a part's form, orientation and location. They are often used therefore without the need for concentricity or co-axiality tolerances and are used primarily on rotational parts but can be applied to any rotational feature. Anyone with knowledge of lathework and dial gauges will readily understand the concept of these. We imagine a part being spun around a central axis and a dial gauge being used to measure the deviation on the surfaces of the part. These surfaces may be **cylindrical, conical, curved profile** or **end faces**. Figure 9.1 shows a set-up where dial gauges are set on cylindrical, conical and planar faces of a part. Run-out tolerances control the maximum deviation that is allowed on the dial; the difference between the minimum and maximum heights is observed. Of course, there are other ways of making these measurements with digital depth meters, CMMs and so on, but it is useful, and often practical, to use dial gauges to conceptualize and measure run-out.

Run-out tolerances come in two varieties. Circular run-out, or simple '**run-out**', is the deviation allowed between two circles which form the tolerance zone.

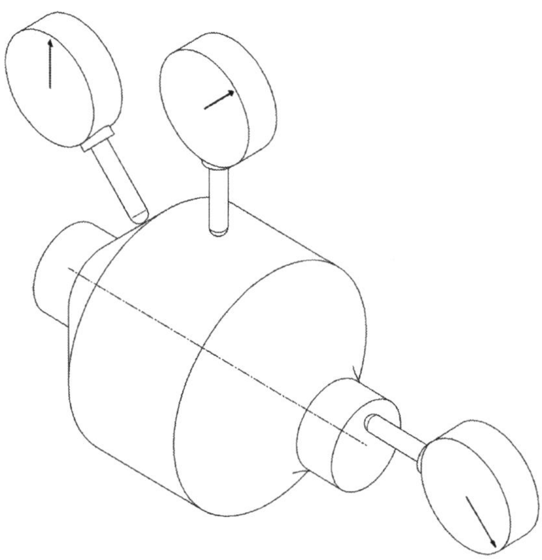

FIGURE 9.1 Dial gauges placed on a rotational part.

DOI: 10.1201/9781003395072-9

Although often characterized as 2D tolerances the circles can lie on a common planar face or on conical or cylindrical faces. '**Total run-out**' is the deviation that is allowed between two surfaces that may be flat, curved, conical or cylindrical and are truly 3D in nature. Table 9.1 shows the symbols that are used for run-out tolerances.

9.2 RUN-OUT

Run-out is defined by the deviation measured normal to a surface that is rotated through a specified angle, most commonly 360° but any angle is allowable. The tolerance zone created is therefore an area between two circles or arcs. Some examples will show part of the range of applications for this ubiquitous tool.

9.2.1 Example—Simple Run-Out

Run-out, when applied to a simple cylindrical surface in relation to a datum axis, is simply the measured deviation between two circles that lie on a plane normal to the datum axis. Figure 9.2 shows how such a tolerance might be represented on a drawing. To validate that the tolerance has been met, a physical set-up might

TABLE 9.1
Symbols for Location Tolerances

Symbol	Name
↗	Run-out
⌰	Total run-out

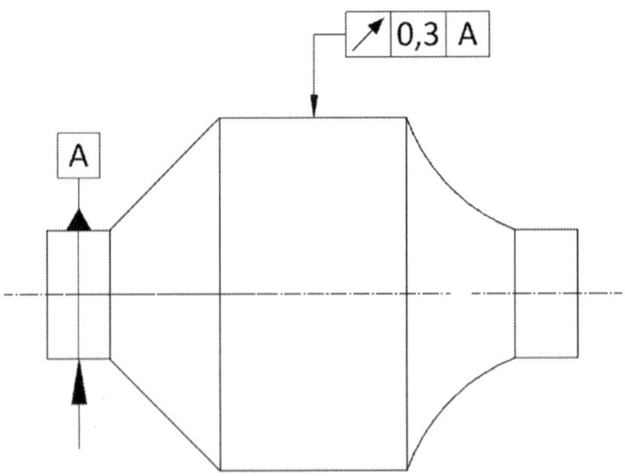

FIGURE 9.2 Simple run-out tolerance.

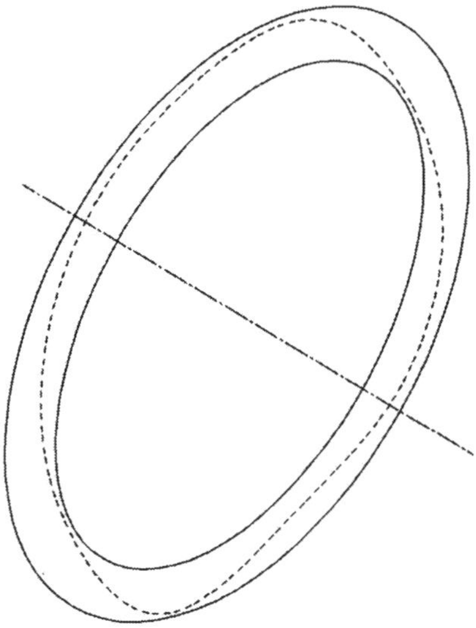

FIGURE 9.3 Tolerance zone for a simple run-out on a cylinder.

involve spinning the part, the axis of rotation forming the datum axis A. A dial gauge is placed so that it is normal for the surface of the cylindrical surface to be measured and the part is turned 360°. If the total deviation on the dial gauge is less than or equal to 0,3 mm, the feature is accepted.

The tolerance zone for a run-out normal to the axis of rotation on a cylindrical surface will be as shown in Figure 9.3. This shows two perfectly concentric and co-planar circles normal to the datum axis with the profile generated from the dial gauge readings. The dial gauge readings (extracted feature) must fall between circles which are concentric and have a radial difference of no more than 0,3 mm.

Note that the independence principle will apply by default. The tolerance applies to any cross-section on the cylinder, but these are independent of each other and the extracted diameters may differ, but as long as they are 0,3 mm apart in each cross-section, they will be accepted.

9.2.2 Example—Common Datum Run-Out

Figure 9.4 shows how a run-out tolerance might be shown on a drawing for a cylindrical surface in relation to a common datum axis. The tolerance zone and validation process will be similar to the previous example.

9.2.3 Example—Cone Run-Out

Figure 9.5 shows how a run-out tolerance might be shown on a drawing for a conical surface in relation to a datum axis A.

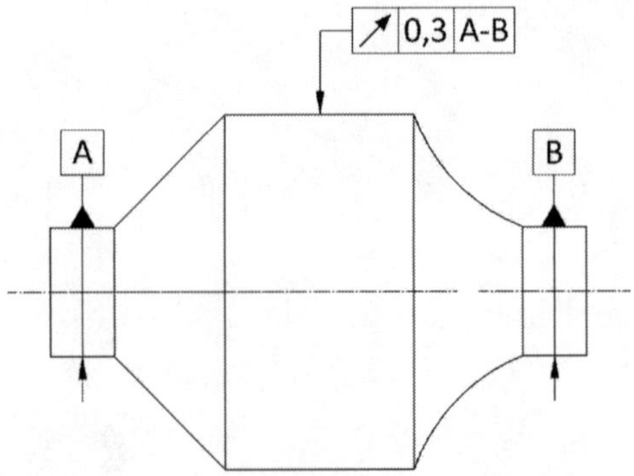

FIGURE 9.4 Common datum run-out tolerance.

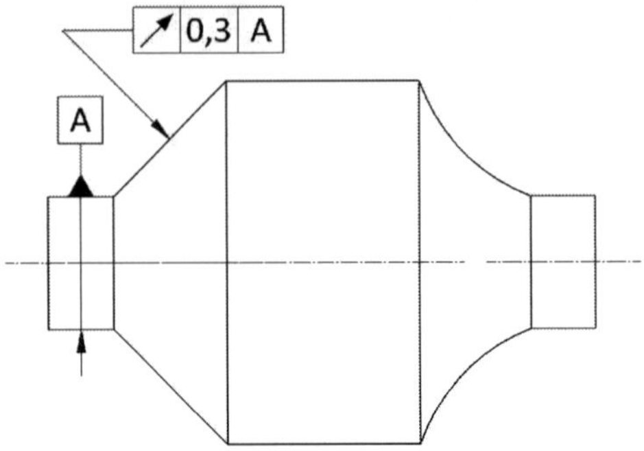

FIGURE 9.5 Cone run-out tolerance.

The tolerance zone for a conical surface is usually at right angles to the cone itself, and the leader line for the tolerance frame should therefore be at right angles to the surface in such cases. The two circles that bound the tolerance zone then lie on different cross-sections of a cone whose angle is normal to the cone being measured. This is shown graphically in Figure 9.6.

9.2.4 Example—Curve Run-Out

Figure 9.7 shows how a run-out tolerance might be shown on a drawing for a curved surface in relation to a datum axis A. As with conical surfaces, the tolerance zone for each independent tolerance zone will be between two circles that lie on different cross-sections of a cone whose angle is normal to the curve at the section being measured (see Figure 9.6). Thus the angle of the tolerance zone will vary along the curve cross-sections.

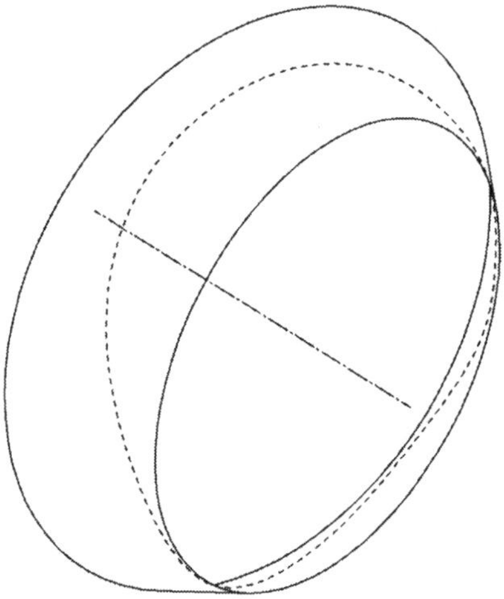

FIGURE 9.6 Tolerance zone for a conical surface.

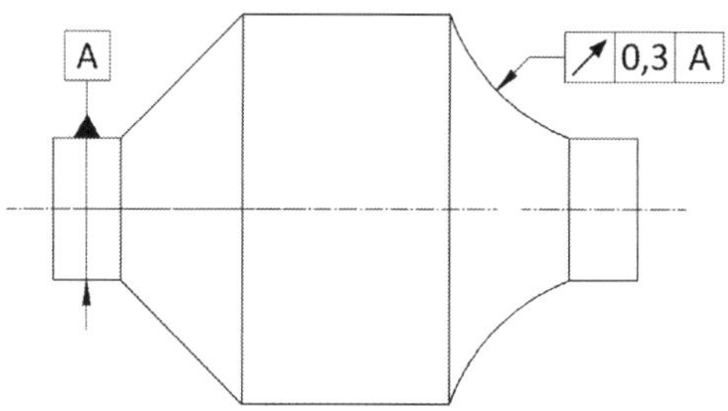

FIGURE 9.7 Curved surface run-out.

9.2.5 Example—Curve Run-Out with Direction Feature

Figure 9.8 shows how a run-out tolerance might be shown on a drawing for a curved surface in relation to a datum axis A, which has a direction indicator that defines the angle of the cone upon which the tolerance zone will lie. This angle is fixed so that in the example shown a dial indicator would always be positioned at 72° to the datum regardless of where on the curve the cross-section is being measured.

9.2.6 Example—End Face Run-Out

Figure 9.9 shows how a run-out tolerance might be shown on a drawing for a flat surface that lies perpendicular to a datum axis A. In such a case a dial gauge would be placed at right angles to the surface and hence parallel with the datum line.

The tolerance zone for such a face will consist of two perfectly concentric circles that lie perpendicular to the datum axis, which are separated by a distance

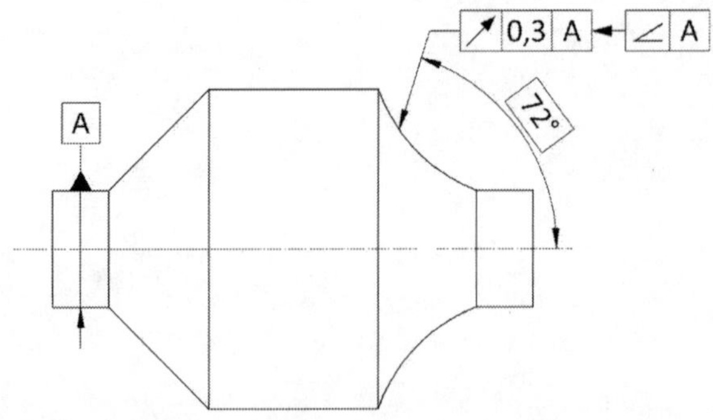

FIGURE 9.8 Curved surface run-out with direction feature.

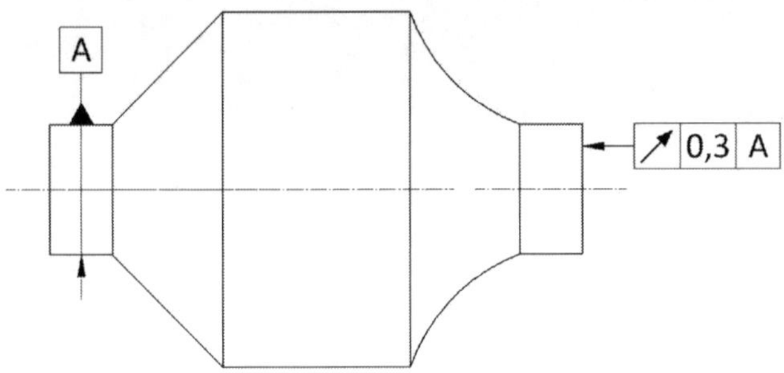

FIGURE 9.9 Run-out on end face.

of 0,3 mm along the axis. This forms a cylindrical face upon which all validation measurements must lie. This is shown graphically in Figure 9.10.

Note that there are an infinite number of measurements that can be made at different diameters, and each is independent from all others. As long as each single measurement at a given radius lies between a 0,3 mm wide boundary, the feature will be acceptable.

9.3 TOTAL RUN-OUT

Total run-out applies to complete surfaces or parts thereof. To be acceptable, an entire feature, or specified part thereof, must fit inside a boundary volume depicted by specified circular faces that are a given distance apart.

9.3.1 Example—End Face Total Run-Out

Figure 9.11 shows how a total run-out tolerance might be shown on a drawing for a flat surface that lies perpendicular to a datum axis A. As in the example shown in Figure 9.7, a dial gauge could be placed at right angles to the surface and hence parallel with the datum line. However, the two circles represent the edges of two planar faces that all measurements at any radius must fit inside. In other words,

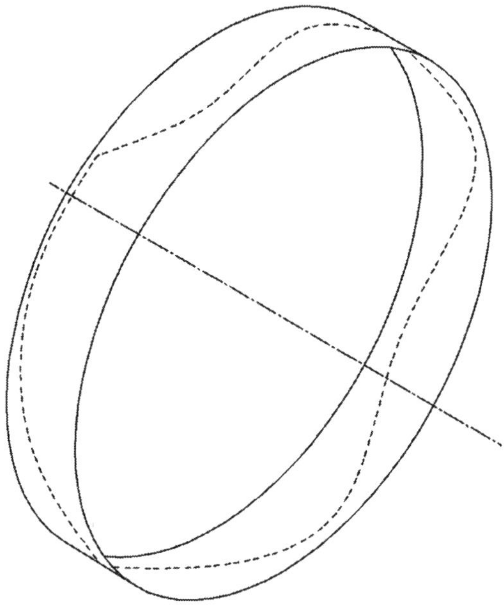

FIGURE 9.10 Tolerance zone for flat face perpendicular to a datum.

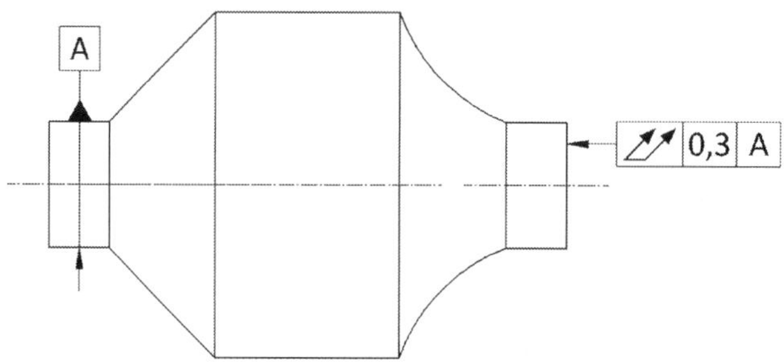

FIGURE 9.11 Total run-out on the end face.

the entire specified surface must fit inside a cylindrical volume bounded by two circular planar faces or discs.

A graphical illustration of an end face encapsulated by its total run-out tolerance zone is shown in Figure 9.12.

9.3.2 Example—Total Run-Out on Cylindrical Surface

Figure 9.13 shows how a total run-out tolerance might be shown on a drawing for a cylindrical surface that lies parallel to a common datum axis A–B. In such a case, the total run-out indicates that the entire specified surface must fit between two cylinders that are coaxial with a radial separation of 0,3 mm.

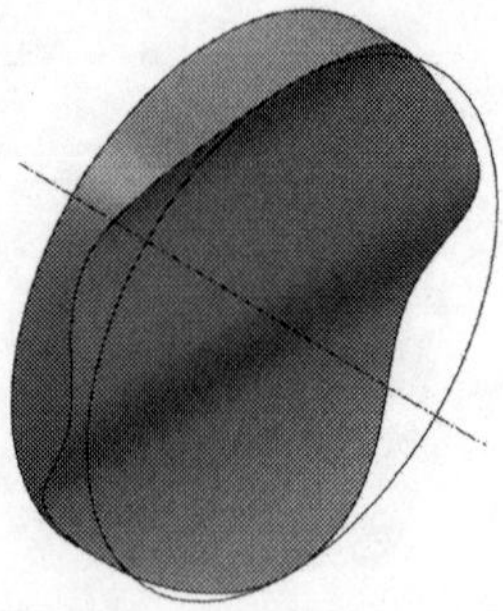

FIGURE 9.12 Tolerance zone for total run-out on end face.

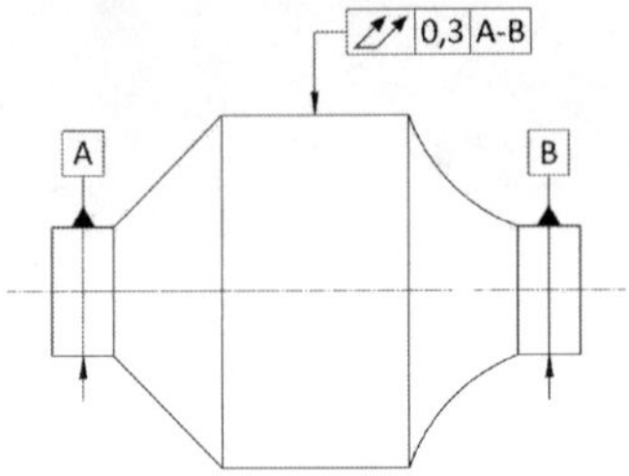

FIGURE 9.13 Total run-out on cylindrical face with common datum.

Figure 9.14 shows with two 2D elevations what the cylindrical tolerance zone (hatched) would look like the previous example.

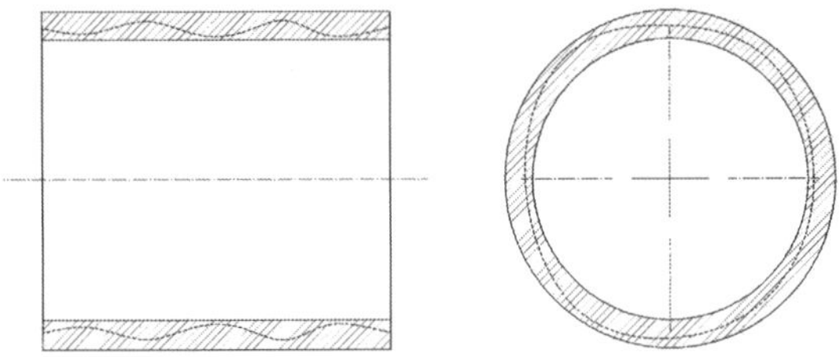

FIGURE 9.14 Tolerance zone (shaded) for total run-out on a cylindrical feature (dotted).

9.4 COMMENTS

We have seen that run-out tolerances are general purpose composite tolerances that are particularly useful for dealing with turned components. At least some of the constraint examples given in the chapter might have been achievable using previously discussed methods or a combination of them, such as the application of roundness, cylindricity, concentricity or coaxiality. Nonetheless, run-out tolerances are widely used because of their convenience in dealing with a large class of manufactured components.

10 Curves and Surfaces

10.1 INTRODUCTION

One of the biggest problems faced by vendors of **CAD** systems is the challenge of developing suitable software for modelling complex **curves** and **surfaces**. Specifying how such elements should be depicted on engineering drawings can itself be difficult, but failure to do so can lead to serious problems downstream from the design office. Engineering standards give some guidance in this area, but the wide range of methods that can be adapted to generate and represent curves and surfaces limits the possibility for standards to be fully definitive. Some understanding of the tools that are commonly available in CAD systems for the construction of complex shapes can help. This chapter provides an introduction to the topic and explains the major techniques that have been developed by mathematicians, engineers and programmers.

If you are designing components with curves and surfaces, you may be blissfully unaware of the mathematics underlying the tools you are using; however, some understanding of how these tools work will help you develop better ways to make more efficient models. It might also enable you to improve ways of communicating design intent to others who are using your drawings or models. If nothing else, as you understand the limitations of the technologies involved, it might help you avoid some of the frustrations that can arise when the CAD system is seemingly refusing to do as instructed!

Complex geometric curves and surfaces are vital in the design of functional structures like wing profiles, ship hulls and vehicle bodywork. They are also very commonly used to create aesthetically pleasing surfaces for consumer goods. It is perhaps not surprising then that many of the major developments in curve and surface technology have occurred in the car industry, where a combination of function and style are vital.

Traditionally engineers drew curves using spline methods. These were techniques that employed pliable materials (such as lead or wooden strips) that could be curved around pins attached to pieces of wood to produce very smooth curves that interpolated control points (i.e. the pins). Splines that pass through chosen control points are therefore called '**interpolation splines**', and those that are merely pulled toward control points are called '**approximation splines**'. Most modern CAD systems will actually carry both an interpolation and approximation definition of any single curve/surface element since both representations can be useful for creation and editing. Let's now consider the methods that have been used to construct many of the most common computer-based spline tools.

DOI: 10.1201/9781003395072-10

10.2 BASIC CURVE GEOMETRY

The mathematics of Cartesian geometry that you possibly learned at school does not lend itself to aspects of some CAD work in the sense that the representations used are often inappropriate. For example, consider the traditional equation of a straight line.

$$y = ax + b$$

This equation is based on the description of a line as a gradient (a) and intercept on the y-axis (b). To describe a line segment we would specify two values of x with which to limit the extent of a given line. This method of describing a line is simple, but it does have drawbacks, not least the fact that it does not cope very well with vertical lines, and therefore a better way to describe line segments is to use parametric descriptions. So our new straight line equation(s), which copes with three dimensions, becomes:

$$x = a1 + b1t$$

$$y = a2 + b2t$$

$$z = a3 + b3t$$

where '*t*' is a common parameter.

The parameter *t* implies time, and it is sometimes useful to imagine resulting lines and curves as being traced out over a time period from, say, 0 to 1, as we shall see. For compactness, it is also useful to represent the previous set of equations in a more brief vector notation as follows,

$$\mathbf{S}(t) = \mathbf{A} + \mathbf{B}t$$

where **S** is the set of coordinate points tracing the curve, **A** = {a1, a2, a3}, and **B** = {b1, b2, b3}.

This method allows us to easily draw robust straight lines. We can set the general nature of a straight line by changing the coefficients ***A*** and ***B***. Straight lines are therefore easy, but now let us consider a simple curve. This can be represented by say a quadratic, cubic, quartic, quintic and so on, polynomial like so:

$$\mathbf{S}(t) = \mathbf{A} + \mathbf{B}t + \mathbf{C}t^2 + \mathbf{D}t^3 + \ldots\ldots\ldots$$

This method of describing curves uses as its 'basis', a series of powers of the independent variable *t* and is therefore known as a **'power basis polynomial'**. This may work well for a straight line where the meaning of the coefficients **A** and **B** is reasonably clear, but the meaning of **C** or **D** is not. If we want to draw higher-order curves, we might be better to use more meaningful design variables. Spline curves are ideally drawn using design variables that are easy to manipulate and that have a clear meaning to the designer; typically these are the positions of a small set of **control points** through or near which a line or curve will pass.

10.3 INTERPOLATION SPLINES

Perhaps the simplest way of describing a curve for a designer would be to use a set of user-defined (mouse or typed) points that would be interpolated. Such splines are analogous to our early example of the techniques used by the shipbuilding industry, fitting wooden strips around pins. This is indeed a useful method, but given a set of $m + 1$ points, we would need a curve of degree m to describe our shape. With many points we end up with very high-degree curves. To plot the curves we have to find interpolating values of the polynomial coefficients, and this involves the solution of potentially large sets of simultaneous equations. These curves can also easily become too 'squiggly' for the designer's liking, and they offer no local control. Moving a control point, even a little, can produce large movements of the curve at positions far removed from the point being edited. In practice these problems are usually overcome using only, say, cubic curve segments which are joined or '**knotted**' together to form piecewise curves that can interpolate many points. So, typically, we create long complex curves by joining together several smaller curve sections. We'll look at how we do the joining later, but first we'll consider some common interpolation techniques that can be used to create the curve segments.

The simplest method of making piecewise curves uses **Lagrangian** interpolation methods to fit curves (typically cubic, degree $m = 3$) through $m + 1$ (typically therefore 4) control points. This involves solving a set of four simultaneous equations. A second common method of developing interpolation splines is by the use of what is known as **Hermite** interpolation. Again, this involves design variables describing control point locations which are additionally associated with a direction vector representing a tangent to the curve at the control point location. Graphical depictions of how the two methods are seen by the user of a typical CAD system are shown in Figure 10.1, where the curves can be manipulated by moving the points or vectors. Both of these allow local control over piecewise curves since any action taken at a control point will only affect the curve segment(s) that reference it.

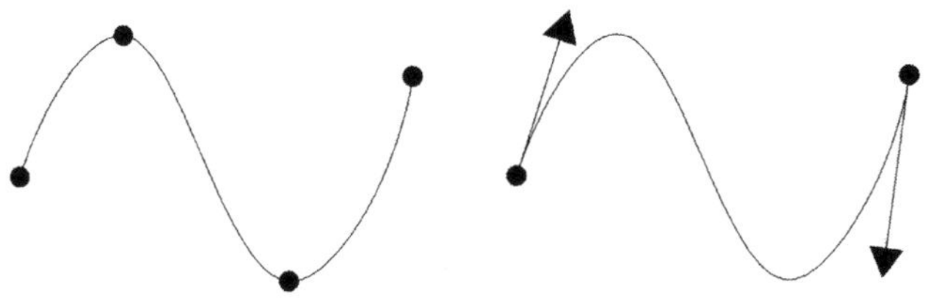

FIGURE 10.1 Lagrangian and Hermite interpolation.

10.4 APPROXIMATION SPLINES

There are several methods that have been developed to produce approximation splines, for example, **Bézier**, **B-spline** (uniform, non-uniform or periodic), **Beta-spline** and **NURBS** (non-uniform, rational B-splines). We will now try to look at the most basic of these to provide an introduction.

10.4.1 Bézier Curves

The first application of approximation splines is credited to the French mathematician **Pierre Bézier**, who worked for the Renault motor company in the 1950s and 1960s. He proposed a method of drawing curves that would provide an intuitive feel for automotive body designers, allowing shapes to be manipulated by 'pulling' on the curves. These curves interpolated end points but approximated intermediate points. This, it is argued, makes the curves easy to position and join together and simple to edit in terms of shape.

A Bézier curve for a simple cubic curve (degree 3) will have four control points as we might expect and might look like the curve shown in Figure 10.2.

Perhaps the easiest way to understand how a Bézier curve might be constructed is to use what is known as the de Casteljau construction. **Paul de Casteljau** was a contemporary of Pierre Bézier and worked for the Citroen company from the 1950s onwards. The work of these two gentlemen has contributed greatly to the mathematical description of curves and surfaces as well as allowing Renault and Citroen to produce some beautiful cars in the 1960s and 1970s. Paul de Casteljau's algorithm is based on a physical analogy of points travelling in straight lines along tracks.

Consider a simple quadratic curve with three control points (P_0, P_1, P_2) as shown in Figure 10.3. Using our time analogy we imagine points travelling along lines from time $t = 0$ to time $t = 1$. At time $t = 0$, a point (P_0') sets off from P_0 travelling along a straight line towards P_1 arriving there at time $t = 1$. Simultaneously a second point (P_1') leaves P_1 ($t = 0$) and travels in a straight line towards P_2, arriving there at

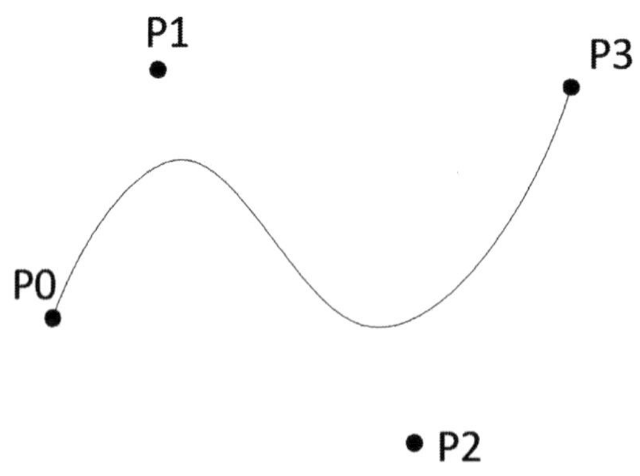

FIGURE 10.2 Cubic Bézier curve.

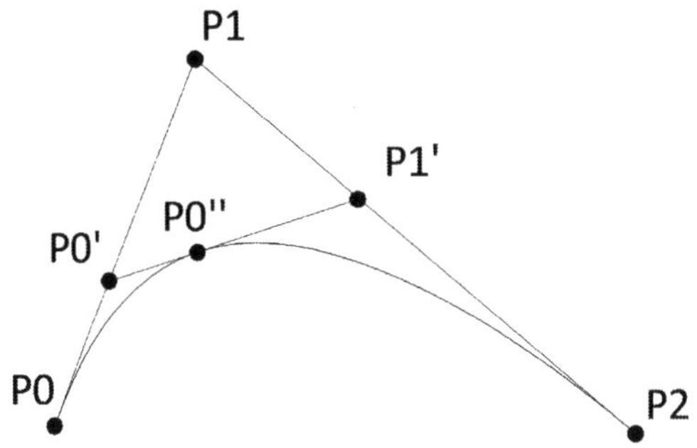

FIGURE 10.3 de Casteljau construction of a Bézier curve.

time $t = 1$. Also, a third point (P_0'') sets off from point P_0' at $t = 0$ and travels along what is now a moving and stretching straight line towards P_1', arriving there at time $t = 1$. The path travelled by point P_0'' follows a quadratic Bézier curve.

Note that the shape defined by the lines between the control points is called the control polygon, in the example in Figure 10.3, a triangle.

We can generate some simple linear equations to allow us to explain the statement mathematically, as follows:

$$\mathbf{P}_0' = (1-t)\mathbf{P}_0 + t\mathbf{P}_1$$

$$\mathbf{P}_1' = (1-t)\mathbf{P}_1 + t\mathbf{P}_2$$

$$\mathbf{P}_0'' = (1-t)\mathbf{P}_0' + t\mathbf{P}_1'$$

By substitution we get, $\boldsymbol{S}(\boldsymbol{t}) = \mathbf{P}_0'' = (1-t)^2\mathbf{P}_0 + 2(1-t)t\mathbf{P}_1 + t^2\mathbf{P}_2$

We can use this approach to generate cubic or other degree curves in the same manner. Generalizing the previous equation for a curve of degree d we get:

$$\mathbf{S}(t) = \sum_{i=0}^{d} B_{i,d}(t)\mathbf{P}_i$$

where $B_{i,d} = \begin{pmatrix} d \\ i \end{pmatrix} t^i (1-t)^{d-i}$ and $\begin{pmatrix} d \\ i \end{pmatrix}$ are the binomial coefficients (1–1 for linear, 1–2–1 for quadratic, 1–3–3–1 for cubic and so on). Thus, the basis of a quadratic polynomial in this form is not a power series but a set of functions as follows:

$$(1-t)^2$$

$$2(1-t)t$$

$$t^2$$

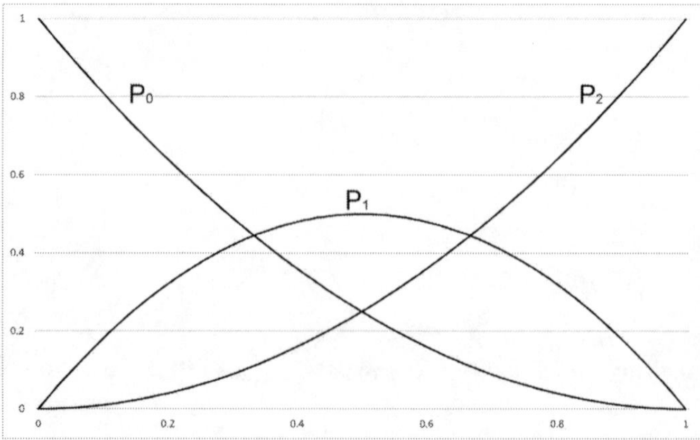

FIGURE 10.4 Blending functions for quadratic Bézier curve.

These functions are known as the blending functions or basis functions and control the amount of 'pull' from each control point at any value of t along the curve. We can plot the blending functions graphically, as shown in Figure 10.4.

The Bézier curve passes through its first and last control points. This property derives from the blending functions since at the endpoints when $t = 0$ the blending function for P_0 is 1 while the others are zero. When $t = 1$ then the blending function for P_2 is one and the others are zero.

It is possible to create higher-order Bézier curves to deal with more complex shapes, and CAD systems will normally allow some adjustment of a curve's degree to allow, say degree 3, cubic curves, or degree 4, quartic curves, and even degree 5, quintic curves. However, as previously discussed, higher-degree curves can become unwieldy and expensive to compute. Also, most modellers like to be able to edit curves using control points that exert local control. One way to achieve this is to join low-degree curve segments together, forming a piecewise composite curve. We can now consider how this might be achieved.

10.4.2 Continuity

The notion of **continuity** refers to the nature of the desired join between curves that are used to make up a piecewise curve, a total curve made up of several individual curves or segments. We can control the contact and smoothness of joins that are used to make up a piecewise curve. Note that in the following examples the author has created a discrete model, and each curve is plotted at 100 individual points along its length. In theory, we can plot any number of points, but being able to see the gaps between curve points will allow the reader to appreciate the notion of velocity along the curve. Consider the two curves shown in Figure 10.5. These clearly depict two curves that are not joined and have zero continuity.

To join the two curves we simply have to ensure that the last control point on the first curve (C1P3) will coincide with the first point on the second curve (C2P0). We can see this graphically in Figure 10.6. These curves are said to have **geometric continuity G0**.

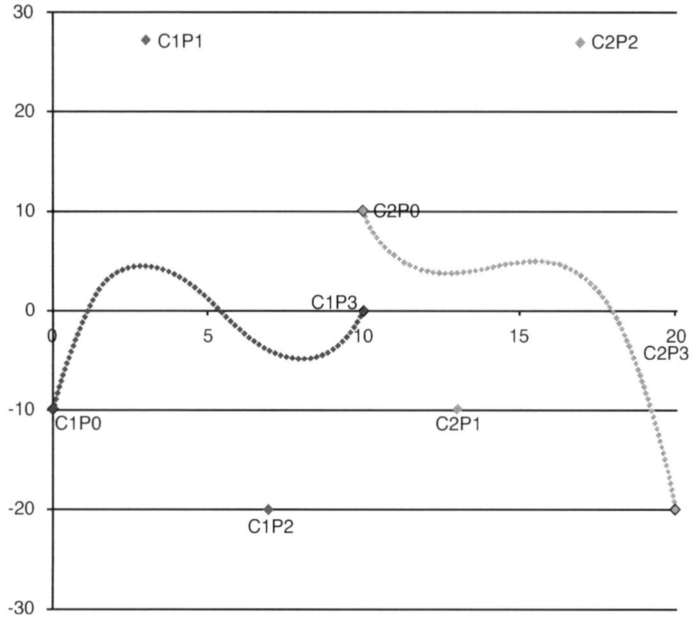

FIGURE 10.5 Two curves with no continuity.

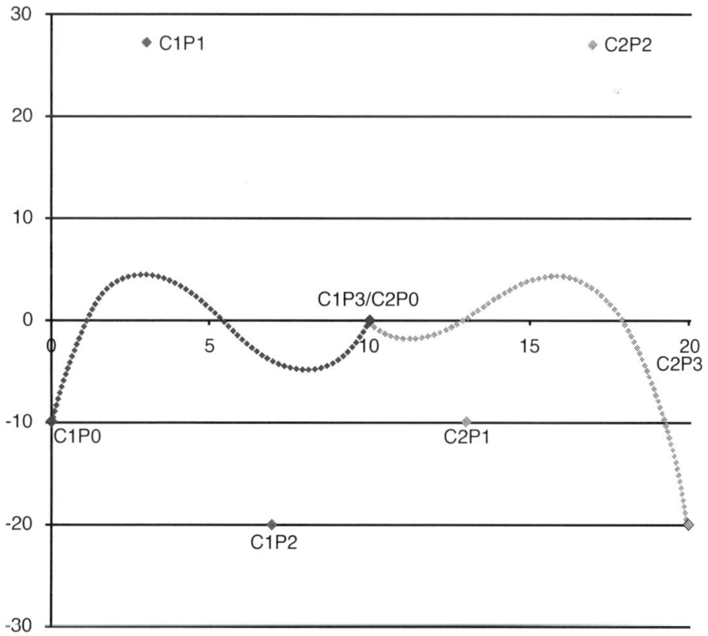

FIGURE 10.6 Two curves with G0 continuity (touching).

If we wish to obtain a smoother join between the curves we remember that the control polygon for a Bézier curve has edges that are tangent to the curve at its ends. So, to ensure a smoother join we can impose tangent continuity between the curves by ensuring the last control point of curve 1 (C1P3) is coincident with the first control point of curve 1 (C2P0) and additionally ensure that the penultimate control point of curve 1 (C1P2), the last control point of curve 1 (C1P3), the first point of curve 2 (C2P0) and the second point on curve 2 (C2P1) all lie along a straight line as shown in Figure 10.7.

Note that this ensures geometric continuity **G1**. The two curves meet with co-linear tangents and hence their geometric first derivatives (dy/dx) are the same at the joint.

If we desire even smother **G2** transition between the curves we can impose equivalence of the second derivatives such that the rate of change of the tangents, or curvature, at the join is equal, as shown in Figure 10.8.

Note that the continuity in the examples relates to geometric continuity (G0, G1, G2 and so on), and this is commonly offered in many CAD systems. If full parametric continuity is desired (depicted C0, C1, C2 and so on), we have to ensure equivalence of the full parametric derivatives (dx/dt, dy/dt and so on). For our points on the discrete graphs this would mean that the curves would have the same sloping tangents at the joins as before and, additionally, that the spacing between the graph points would be the same for each curve at the joins; thus, the 'velocity' of the curves would match rather than simply the direction.

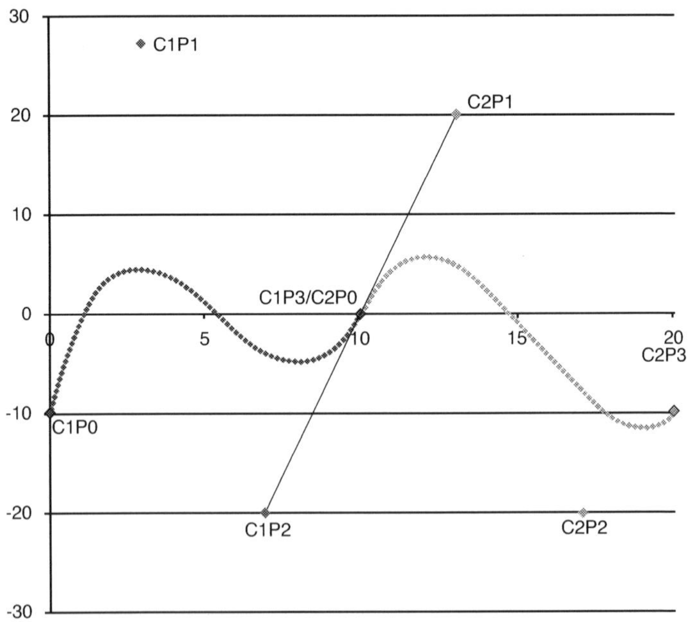

FIGURE 10.7 Two curves with G1 continuity (common tangents).

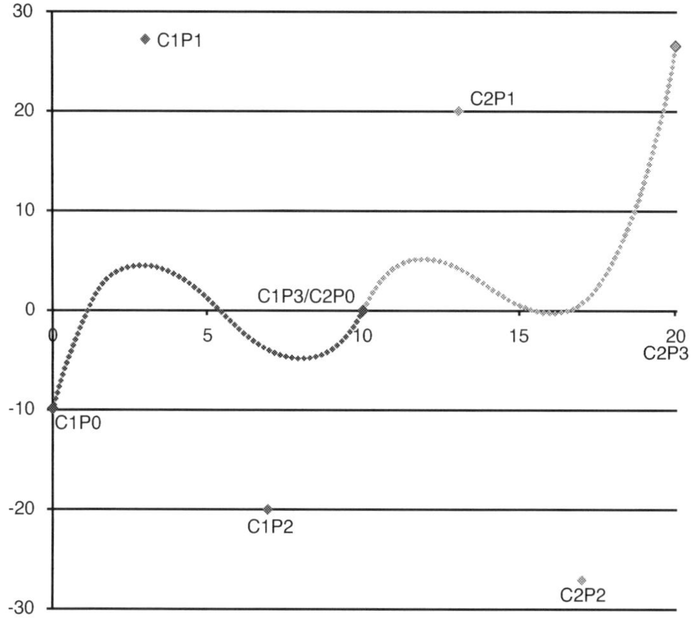

FIGURE 10.8 Two curves with G2 continuity (equal geometric second derivatives).

10.4.3 Benefits of Bézier Curves

Bézier curves are generally reasonably easy to understand, and they have been used successfully in CAD systems for many years.

A curve defined by its control points is invariant over common transformations; for example, it can be infinitely scaled without any loss of form. Transformations such as spinning around axes and so on require only that the control points be transformed, and the new curve projection can be calculated using the predefined algorithm (Bézier with a known degree and continuity).

The previous point also means that the curve geometry can be stored with few storage requirements as only the coordinates of the control points and specification of degree and continuity need to be kept. This makes it easy to transfer curve data between systems with no loss of precision at any scale. A common example of this in practice is the use of true-type fonts, which use Bézier specifications for the shape of the letters so that these can be stored efficiently, transferred and or printed without any loss of fidelity at any scale.

10.5 B-SPLINES

B-splines are, in many ways, similar to Bézier curves and may be thought of as Bézier curves that are '**knotted**' together and overlap. Many complex-looking B-spline curves can be reduced algebraically to their Bézier equivalents. Because of the flexibility in the knotting methods employed in B-spline formulations (more later), they are also naturally **non-uniform**. A B-spline is also naturally a piecewise curve where the method of joining curve segments is already formulated in the equations used to create the graphics.

When we formulate a B-spline curve, we use slightly different basis functions from those shown for the Bézier in Figure 10.4.

The major advantage of B-splines over Bézier curves is that we achieve continuity in piecewise curves by sharing control points within the formulation. The number of points shared is dependent on the degree of continuity required. This means that local changes to the curve are possible.

A B-spline curve can be defined in a similar way to a Bézier curve, that is:

$$\mathbf{S}(t) = \sum_{i=0}^{n} \mathbf{B}_{i,k}(t)\mathbf{P}_i$$

where $\mathbf{P}_i$ are the $n + 1$ control points, the degree of the polynomial segments is $k - 1$, and $\mathbf{B}_{i,k}$ are the normalized B-spline blending functions.

This looks very similar to the Bézier formulation we first presented; however, the blending functions are quite different. These are defined by a recursive formula (known as the Cox-deBoor formulation):

$$\textit{if } t_i \le t \le t_{i+1} \text{ then } \mathbf{B}_{i,1}(t) = 1 \text{ else } \mathbf{B}_{i,1}(t) = 0$$

$$\mathbf{B}_{i,k}(t) = \frac{t - t_i}{t_{i+k-1}}\mathbf{B}_{i,k-1}(t) + \frac{t_{i+k} - t}{t_{i+k} - t_{i+1}}\mathbf{B}_{i+1,k-1}(t)$$

B-splines are essentially a number of curve segments which overlap each other. Commonly each curve segment may be quadratic or cubic or increasingly even quartic or quantic, particularly in applications such as automotive body design. The range of parameter t is divided into $n + k$ subintervals, with each blending function being defined across k subintervals. To control the extent of the overlap and the positions of the joints between segments and their continuity, it is necessary to define what is known as a **knot vector**. This is simply a list of the $n + k + 1$ values in parametric space (along t) that define the span positions over which the spline will be evaluated. In the class of B-splines known as 'uniform' an even spacing is chosen between the knots, and these may be normalized or represented by integers, for example, {0, 0.17, 0.33, 0.5, 0.67, 0.83, 1} or {0,1,2,3,4,5,6}. For $n + 1$ control points, the curve is described by $n + 1$ blending functions and the curve will have continuity of C^{k-2}.

10.5.1 Example—Uniform Quadratic B-Splines

A simple example should make the formulation clear. Suppose we wish to define a uniform quadratic B-spline with four control points, we select $n = k = 3$. The knot vector (we'll use an integer one) will be:

$$\{0,\ 1,\ 2,\ 3,\ 4,\ 5,\ 6\}$$

So, the range of the parameter t is from 0 to 6 with 6 subintervals. Each blending function spans $k = 3$ subintervals. The blending functions can be calculated as follows:

$$\text{for } 0 \le t < 1, \mathbf{B}_{0,3}(t) = \frac{1}{2}t^2$$

$$\text{for } 1 \le t < 2\, \mathbf{B}_{0,3}(t) = \frac{1}{2}t(2-t) + \frac{1}{2}(t-1)(3-t)$$

$$\text{for } 2 \le t < 3\, \mathbf{B}_{0,3}(t) = \frac{1}{2}(3-t)^2$$

The remaining blending functions can be described in a similar way and by simply shifting the starting position by one and substituting t by $t - 1$, that is:

$$\text{for } 1 \le t < 2, \mathbf{B}_{1,3}(t) = \frac{1}{2}(t-1)^2$$

$$\text{for } 2 \le t < 3\, \mathbf{B}_{1,3}(t) = \frac{1}{2}(t-1)(3-t) + \frac{1}{2}(t-2)(4-t)$$

$$\text{for } 3 \le t < 4\, \mathbf{B}_{1,3}(t) = \frac{1}{2}(4-t)^2$$

Similarly,

$$\text{for } 2 \le t < 3, \mathbf{B}_{2,3}(t) = \frac{1}{2}(t-2)^2$$

$$\text{for } 3 \le t < 4\, \mathbf{B}_{2,3}(t) = \frac{1}{2}(t-2)(4-t) + \frac{1}{2}(t-3)(5-t)$$

$$\text{for } 4 \le t < 5\, \mathbf{B}_{2,3}(t) = \frac{1}{2}(5-t)^2$$

And

$$\text{for } 3 \le t < 4, \mathbf{B}_{3,3}(t) = \frac{1}{2}(t-3)^2$$

$$\text{for } 4 \le t < 5\, \mathbf{B}_{3,3}(t) = \frac{1}{2}(t-3)(5-t) + \frac{1}{2}(t-4)(6-t)$$

$$\text{for } 5 \le t < 6\, \mathbf{B}_{3,3}(t) = \frac{1}{2}(6-t)^2$$

And the blending functions would look like the curves shown in Figure 10.9.

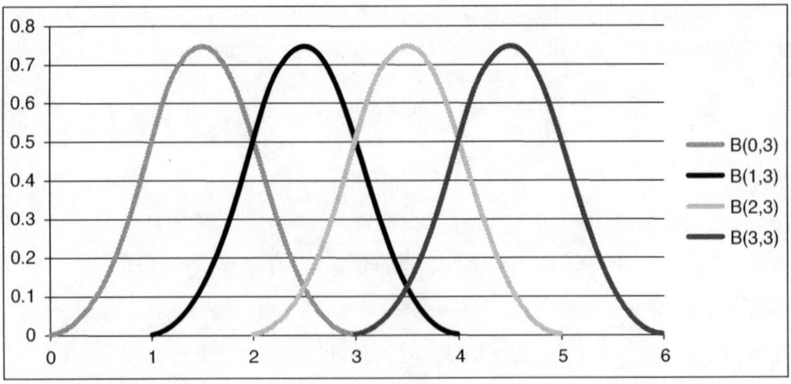

FIGURE 10.9 B-spline blending functions.

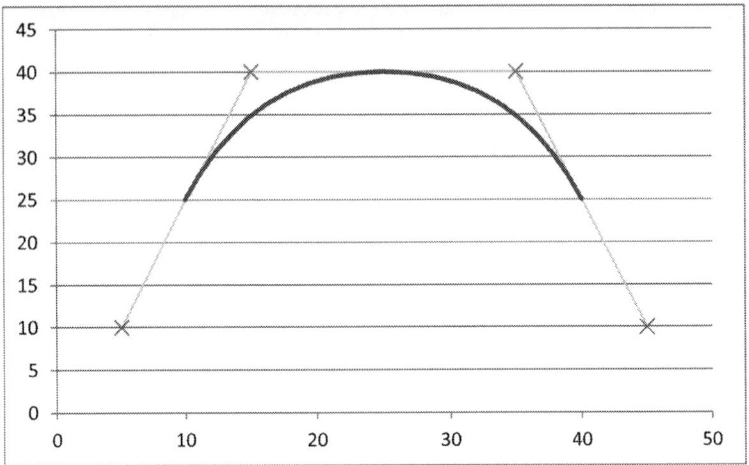

FIGURE 10.10 B-spline made up of quadratic segments.

As you can imagine we could draw these curves ad infinitum for desired increasing numbers of control points. We only draw the actual curve in this case in the regions where it is fully defined (i.e. where the sum of the blending functions is 1) and this is between $t = 2$ and 4.

Choosing control point coordinates (5,10), (15,40), (35,40) and (45,40) would yield the curve shown in Figure 10.10.

10.5.2 More Complex B-Splines

We can make the curves more elaborate in a number of ways. Commonly CAD vendors will set the knot vector so that several joints are superimposed; for example, {0, 0, 0, 1, 1, 1} would appear as having three control points, but the curve would pass through the first and last, similar to the Bézier curve. Going further, complex non-uniform spacing can be used to adjust the continuity along the curve. A common form of **rational** B-spline is developed using what is known as NURBS.

10.5.3 NURBS

NURBS are now used in a number of, if not most, major CAD systems. One of the problems with basic Bézier and B-spline curve drawing methods is that they do not readily make good approximations to common conical curves such as circles, ellipses, parabolas and hyperbolas. Using rational curves we can achieve exact representations of conics. A rational form of B-spline can use the following formulation:

$$\mathbf{S}(t) = \frac{\sum_{i=0}^{n} w_i \mathbf{B}_{i,k}(t)\mathbf{P}_i}{\sum_{i=0}^{n} w_i \mathbf{B}_{i,k}(t)}$$

where w_i are a set of weights applied to the individual control points. The higher the value assigned to a weight, the more its control point will pull the curve towards it. Note that we can, in a similar way, make rational Bézier curves too.

10.6 CURVES IN PRACTICE

There are many ways to develop curve shapes in design. We might start with some governing engineering equations, sculpted surfaces, pieces of string with weights attached, bending pieces of material around formers, taking results from optimization applications based on finite element analyses and so on, but if these curves are to be represented, stored and communicated with the use of CAD systems, then we are faced with the problems of finding the best representation to meet the design purpose.

10.6.1 Conics

A very common requirement when using CAD systems is for the modelling of **conic** sections. The representation of these may be pre-programmed in the CAD system as conics, with their governing equations, or they may use interpolation and/or approximation splines as outlined earlier in this chapter. To generate a circle, for example, with Bézier curves, we can use four individual quadratic curves and join these with G1 continuity such that each curve represents a quadrant of the circle. We can represent this with the specification of a quadratic Bézier with $(4 \times 3) - 12$ control points. Given the restrictions imposed by G1 continuity, we can reduce this to eight control points since the joins will be sharing points.

Using the more general NURBS formulation, we can specify a quadratic formulation with knot vector {0, 0, 0, 1, 1, 1} (which is the same as a quadratic Bézier), and for a single curve section we expect three control points. We can choose the values of the weights to get conics as follows:

$w_0 = 1$
$w_1 > 1$—hyperbola
$w_1 = 0{,}5$—parabola
$w_1 < 0{,}5$—ellipse
$w_1 = 0$—straight line
$w_2 = 1$

The result of all this is that NURBS make a convenient way to represent curves. This is useful to the programmer, but a CAD user may not know what kind of curve formulation is being used to generate and store their geometry. In most cases the formulations are hidden from the user who will be presented with free-form curves depicted by interpolation and/or approximation splines, whereby control points can be specified through a mouse, a graphics tablet, by typing their coordinates directly or by reading values from tables such as spreadsheets. The user might additionally have some further control over the locality of editing curves and the sharpness of cusp points and so on, but these are usually presented as part of the CAD interface and the underlying spline formulations are hidden.

10.6.2 Power Series Curves

Some CAD systems allow the user to enter functions that will generate a specified geometry. These can be formulated as traditional power series equations ($y = ax^2 + bx + c$ and so on), and the CAD system will transform the results of these into the spline formats.

10.6.3 NACA Profiles

In many application areas, for example, wing profiles, curves are specified with geometric functions that may be encoded with values to generate a limited set of possible shapes. Perhaps the best known of these are **NACA profiles**. These consist of a number of different parametric formulations which generate the cross-sections of wings based on engineering data gleaned from simulations or wind tunnel testing of various aerodynamic properties, such as lift or drag, for example.

The first of these that was developed and perhaps most commonly used is the four-digit series of profiles. This is based on three parameters, the maximum thickness of the profile (given as a percentage of the chord length), the camber (given as a percentage of the chord length) and the location of the maximum camber (given as a tenth of the percentage of the distance of the chord length from the leading edge or nose). Figure 10.11 shows a parametrized wing profile as depicted by the NACA 4 series.

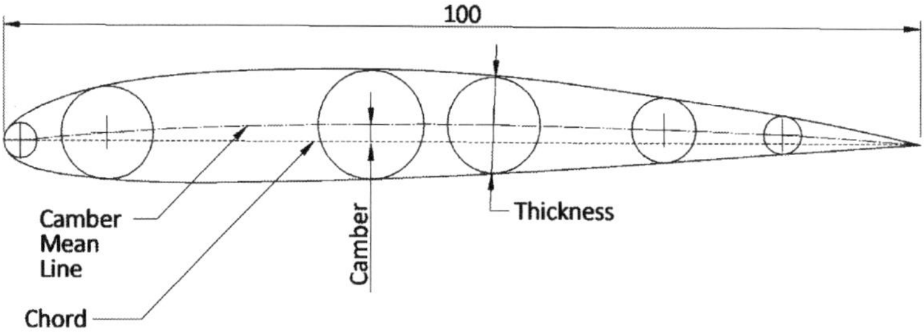

FIGURE 10.11 Profile of a wing depicted by a NACA four-digit series.

An example of a common wing profile is the NACA 4 series 2412. This has a 2% camber (first digit) located 40% back from the leading edge (10 times the second digit) and a maximum thickness of 12% of the chord length (digits 3 and 4).

To plot the exact coordinates of a four-series wing camber, we can use the following equations:

$$y_c = \frac{m}{p^2}\left(2px = x^2\right) for\, 0 \le x \le p$$

$$y_c = \frac{m}{(1-p)^2}\left[\left((1-2p)+2px-x^2\right)\right] for\; p \le x \le 1$$

where

x = ordinates along the chord from 0 to c (the chord length, 100 in Figure 10.11)
m = maximum camber (2 for a 2412 foil)
p = position of maximum camber along the chord (4/10 for a 2412 foil).

The thickness of the foil is measured normally to the camber so the coordinates for the upper and lower profiles become:

$$x_U = x - y_t \sin\varnothing, y_U = y_c + y_t \cos\varnothing$$

$$x_L = x + y_t \sin\varnothing, y_L = y_c - y_t \cos\varnothing$$

where

$$y_t = 5t\left[0{,}2969 - 0{,}1260\sqrt{x} - 0{,}3516x^2 + 0{,}2843x^3 - 0{,}1015x^4\right]$$

$$\varnothing = arctan\frac{dy_c}{dx}$$

$$\frac{dy_c}{dx} = \frac{2m}{p^2}(p-x)\, for\, 0 \le x \le p$$

and

$$\frac{dy_c}{dx} = \frac{2m}{(1-p)^2}(p-x)\ for\ p \le x \le 1$$

So, the previous equations can be used to plot exact coordinates for a four-digit series profile; however, they do not yield a zero thickness at the trailing edge and so an adjusting form has to be applied there, normally a radius whose value depends on FE simulation requirements and/or manufacturing constraints. There are a number of other profiles in the NACA series, for example, five-, six-, seven- and eight-digit series, but these require the use of look-up tables to develop their coordinates. There are a number of specialist computer programs and online resources which can plot a wide variety of wing profile types, including NACA and many others. Some of these will export data directly to spreadsheets or into CAD formats; however, these are usually in the form of equally spaced sample points around the profile, as either Cartesian or polar coordinates.

10.7 GOOD PRACTICE FOR CURVE PROFILES

Plotting wing profiles, or any profile, with a large number of interpolation points is usually a less-than-ideal way to represent complex curves. As was discussed previously, many points can lead to either high-degree curves that are difficult to manipulate or to the use of very many curve sections. In most cases, a CAD curve (spline) needs to be fitted to this data, and it is left to the designer to make judgements about how accurate the fitted curve needs to be.

Generally, it is a good idea to adhere to the following.

Use as few control or interpolation points as are needed to depict form. Fewer is usually better as long as the curve gives an accurate representation of the desired geometry. Many CAD systems will offer 'simplify curve' or similar commands to semi-automate the process of reducing the number of control or interpolation points.

Keep the degree of the curve to a reasonably low value if possible. High-degree curves are difficult to not only compute but also manipulate if they have very demanding continuity constraints. For most purposes quadratic or conic curves are sufficient and offer a good balance between simplicity and the need to resolve constraints imposed by interpolation and tangency. Higher-degree curves may be necessary for very complex applications and in particular for aesthetic reasons, where continuity constraints will also be demanding, for example, for car bodies.

Wherever possible, keep the curve representation as close to its design requirement and generation method as possible. CAD systems which offer equation-driven curves are very useful for this purpose.

10.8 CONVEYING SHAPE

Once a CAD system has been used to generate a curve the designer is often left to decide how best to convey the dimensions to another application, for example, for FE analysis or for manufacturing and validation. There are a number of ways this can be achieved as follows.

10.8.1 Patterns

Perhaps the simplest way to convey the shape of a curve is to print it full-size directly onto a sheet or sheets that can be used as a pattern with which to cut the shape. This may be necessary but will of course incur some loss of fidelity.

10.8.2 Coordinate Values

The **ISO 129–1** standard gives examples of how profiles might be dimensioned in engineering drawings and these are based on placing interpolation ordinates on a profile at equispaced intervals along a guideline, or, on using coordinates (as shown in Chapter 2). When adhering to ISO GPS standards, these should be dimensioned with ordinate/coordinate positions given by TEDs with an accompanying profile tolerance indicator. Coordinate data can be tabulated for clarity if necessary.

10.8.3 Curve Descriptions

For all the reasons previously discussed representing coordinate data are not always the best option to achieve a high-fidelity curve. Where possible, it may be better to give curated geometry, for example, the radius of the wing leading edge or other key parameters, constraints or equations. If this is done it will be necessary to ensure that whatever or whoever is receiving the data can satisfactorily decode and use it. The receiver of the curve information may require a specific style of representation to enable the programming of numerically controlled manufacturing or metrology equipment.

10.8.4 Direct from CAD

ISO 16792 gives direction on the adoption of CAD data that can be used directly to communicate design information. At the time of writing, there is no universal standard for the transfer of spline data; however, the ISO standards allow for this information to be sent as part of the CAD model files that may accompany drawings. As well as native CAD models, there are a number of neutral file formats that can recognize various curve and surface formats. Some care should be taken to ensure accurate transfer however, and this will be discussed in Chapter 11 which describes some popular file types and their import/export options.

10.9 SURFACES

A major advantage of parametric methods of representing curves is that they are easily generalized to deal with surface representations. Instead of a single dimension parameter t we usually use two curve dimensions for our parameter space and label these as u and v. Both the Bézier curve and the B-spline can be extended to create surface patches. There are many ways to characterize or classify surfaces depending on their geometry or how they might be constructed but strictly speaking a 'surface' can be a plane surface on a solid, however, most of the time when we talk about 'surfaces' we mean the curvy type. Sometimes these are infinitely thin **constructions** and sometimes they are used to 'contain' solids by bounding these. When used as a boundary for a solid, we must give the surface a direction (known as the '**surface normal'**) by defining a simple unit vector that points away from the surface into free space.

Perhaps the simplest surface, from the point of view of constructing it, is a curve that is extruded along a straight line in space as shown in Figure 10.12. We can think of this as a curve which is a function of a parameter u being swept along a second curve which is a function of a parameter v. When that sweep is along a curve that is a simple straight line most CAD systems refer to this as **extruding**.

We can create many types of surface by simply taking curves and manipulating them by extrusion, as we have shown, but also by spinning them (**revolved surfaces**), spanning between curves (**lofting**) translating them along other curves (**sweeping**), using contacting curves (at least G0) to form closed boundaries (**bounded surfaces**), or by combining these, often in ways that support intuitive styles of working. Created surfaces can be used to further construct other geometry and may be trimmed, offset or solidified by thickening, stitching or by replacing existing solid faces. However, underneath all of the commands commonly used in CAD systems, are the means to mathematically describe the surfaces and these are based on the same methods that we used to describe curves, predominantly interpolation splines and approximation splines.

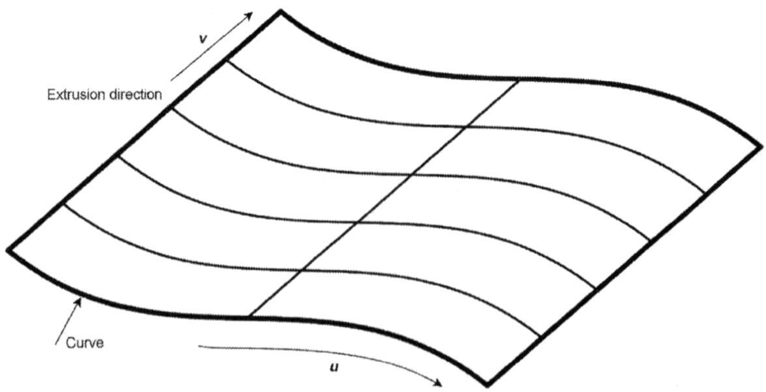

FIGURE 10.12 Simple extruded surface from a curve.

10.10 BÉZIER SURFACE PATCHES

When each point on a Bézier curve is moved along a second Bézier curve at some angle to the first, a Bézier **surface patch** is defined. These 'surface patches' are defined by the network of points that form the control polyhedron in the same way a curve was defined by its control polygon. Many of the properties of the patch are born from those belonging to the Bézier curves.

10.11 B-SPLINE SURFACE PATCHES

We move to B-spline surfaces in the same way as we moved from Bézier curves to surface patches. Importantly, this gives us a surface with which the continuity is independent of the number of control points and can be locally manipulated. The rest of the useful B-spline properties also apply.

10.12 RATIONAL PARAMETRIC SURFACES

Rational parametric surface patches are also analogous to their curve counterparts and are defined by rational versions of the equations of Bézier or B-spline surfaces. This results in surface representations that are capable of modelling spheres and cylinders as well as surfaces based on ellipses, parabolas and hyperbolas.

10.13 NURBS SURFACES

If we implement a B-spline as a rational function, as has been described in sections 10.6 and 10.13, we get NURBS. These form the basic curve and surface geometric representations that are widely used in most high-end and mid-range CAD systems and are also frequently used in file transfer standards. Understanding them therefore is useful in understanding how the basic geometry of designs can be accurately communicated to downstream functions without loss of fidelity, as would be expected from drawings or simple ordinate dimensioned drawings.

10.14 SURFACES IN PRACTICE

Many of the operations to create surfaces in CAD systems can be used in two modes—to create construction surfaces or to create solids bounded by surfaces. CAD systems usually display these differently. Commands that are used to create solids are usually also available in a form that can create a construction surface; however, there are usually many additional commands that can be used to create construction surfaces since there are usually fewer constraints on their form. Often, we build a complex model by making construction surfaces and editing these before using the result to create a solid. However, to build robust models, it is sometimes useful to form solids where possible and use construction surfaces for editing where they are needed. This approach helps the user to spot potential problems (often due to issues at the joints of surfaces) early.

We can consider some further common surface tools.

10.14.1 Lofting

In addition to simple extruded surface patches, a common and simple method of constructing these is to get the CAD system to automatically span between two curves. For example, see Figure 10.13:

10.14.2 Sweeping

A basic sweep involves 'sweeping' or translating one curve along another as shown in Figure 10.14, where the so-called cross-section curve is swept along the rail curve.

Note that we equally sweep the rail curve along the cross-section curve.

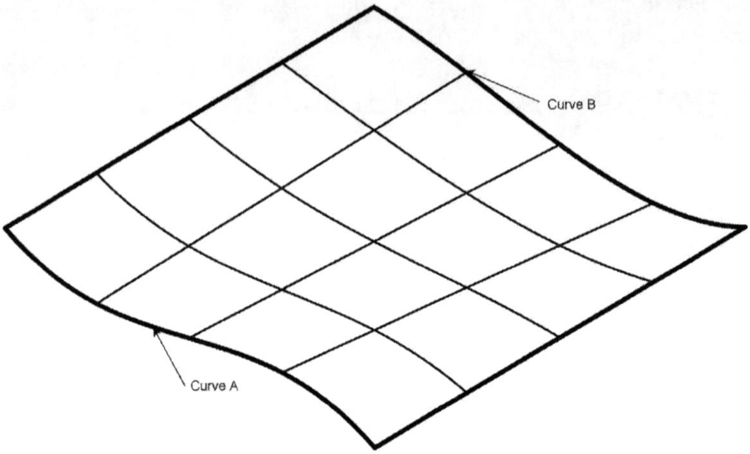

FIGURE 10.13 Basic loft between two curves.

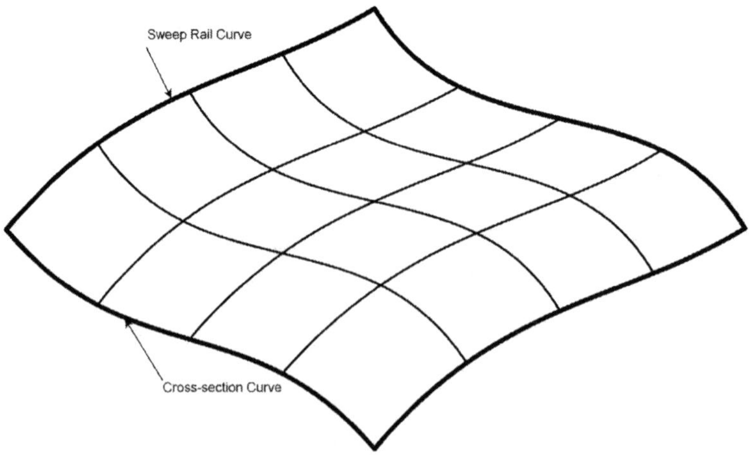

FIGURE 10.14 Simple sweep of cross-section curve along rail curve.

More complex sweeps can be used to create surfaces that can be thought of as combined sweeping and lofting so that we can, for example, sweep one curve, say 'A' (a 'cross-section' curve) along another, say 'B' (a 'guide' curve) while the original curve 'A' is changed continuously as it sweeps until it adopts the shape of a final cross-section curve defined by a final curve, say 'C'. For rectangular type patches we can define two cross-section curves at opposite sides of the patch as well as two rail curves at opposite sides.

Modern CAD systems are likely to offer several options on sweep type commands and offer such multiple cross-section and multiple rail, or 'guide', curves. Commonly, a cross-sectional curve might be swept along a circular path to create a revolved shape.

10.14.3 Bounded Surfaces

A common requirement in a design scenario is to fit a surface patch into a bound region where a number of curves may be joined together to form the boundary. For example, four edge curves (from extruded surfaces) are used to create a boundary as shown in Figure 10.15.

Note that the four curves, labelled 1–4, meet at the corners. However, where a curve section such as that labelled a meets b or where c meets d there is a sharp change of direction and these pairs of curves therefore only join with G0 continuity. We therefore know that we cannot fit a surface patch that can have more than G0 continuity with all four of the surfaces that surround it. Understanding this will save anyone who is starting out designing surfaces a great deal of frustration.

So, we can fit a surface patch to the boundary surfaces with G0 continuity as shown in Figure 10.16.

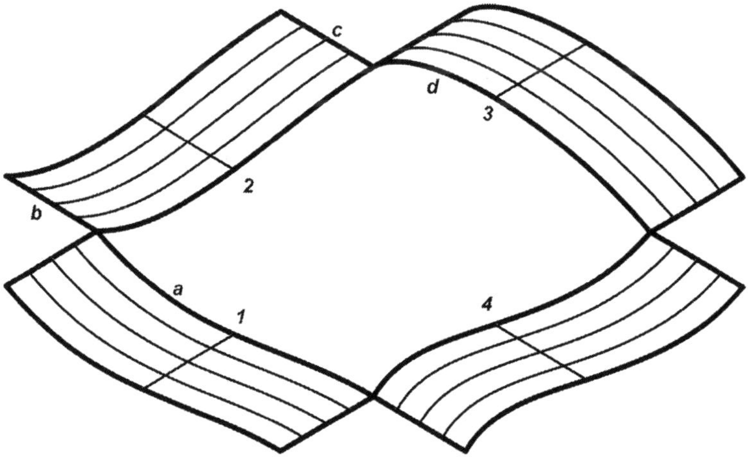

FIGURE 10.15 Four (extruded) surfaces form a boundary.

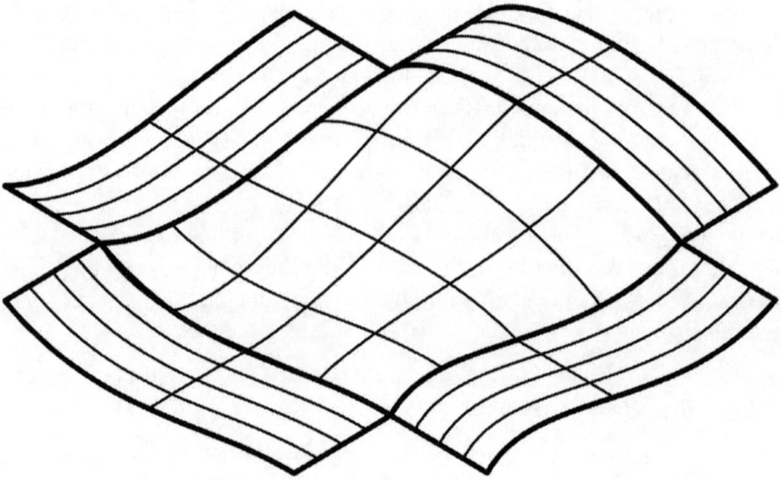

FIGURE 10.16 Surface patch fitted to four extruded surfaces with G0 continuity.

10.15 CONTINUITY OF SURFACES

Figure 10.17 shows two surfaces joined with G0 continuity. This is clear to the observer as the patches touch along a common edge but the contour lines do not match smoothly across the join.

It can be difficult to visualize the continuity between such surfaces and so CAD systems usually have tools to aid the modeller in this area. One such tool uses what is known as 'Zebra Stripes'. Figure 10.18 shows the same curve as in Figure 10.17 but with the zebra stripes showing. Notice how the stripes do not meet up across the join between the surfaces. This indicates G0 continuity.

In Figure 10.19 two surfaces that meet with G1 continuity, but it is difficult to tell if this is the highest level of continuity on the join.

Figure 10.20 shows the same surfaces as in Figure 10.19. The zebra lines match position across the join but they do not do so smoothly. This indicates that the surfaces join with G1 continuity.

In Figure 10.20, two surfaces join with G2 continuity. The join looks smooth but again it is difficult for the observer to know exactly what level of continuity there is.

Figure 10.22 shows the same surfaces as in Figure 10.21. The zebra stripes match position across the join and the join becomes invisible because they also meet smoothly. This indicates that the surfaces join with their tangents aligned and their curvature (second derivatives) match, so the join is G2 continuous.

You may have noticed that some of these surfaces that have been formed in the examples may have been created in multiple ways. For example, the bounded surface in Figure 10.16 may be thought of as a loft with guide curves added or a sweep with multiple guide curves and cross-sections or simply as a bound surface. No matter what commands are actually used these surface patches are normally

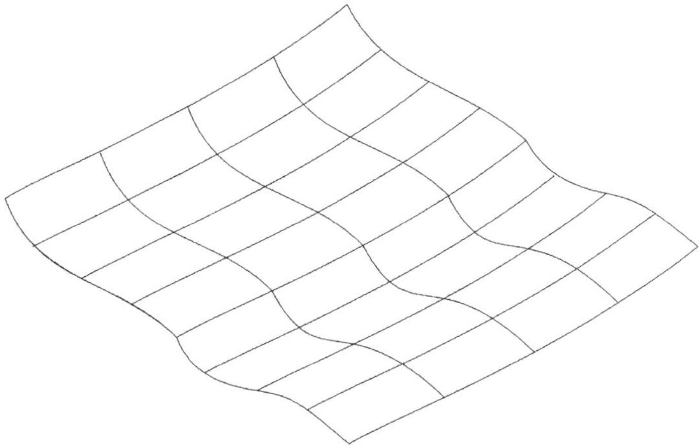

FIGURE 10.17 Two surfaces with G0 continuity.

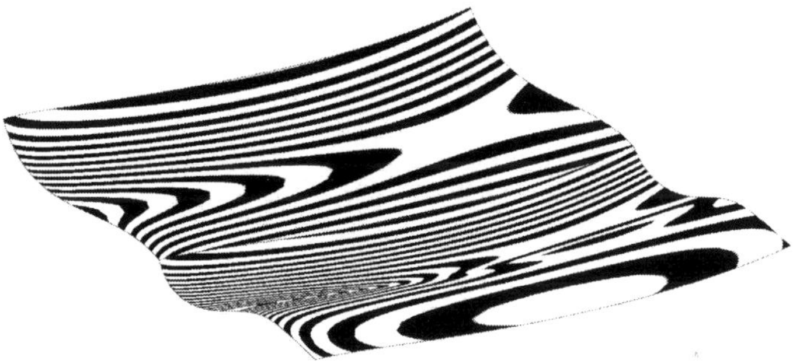

FIGURE 10.18 Zebra analysis of G0 surface join.

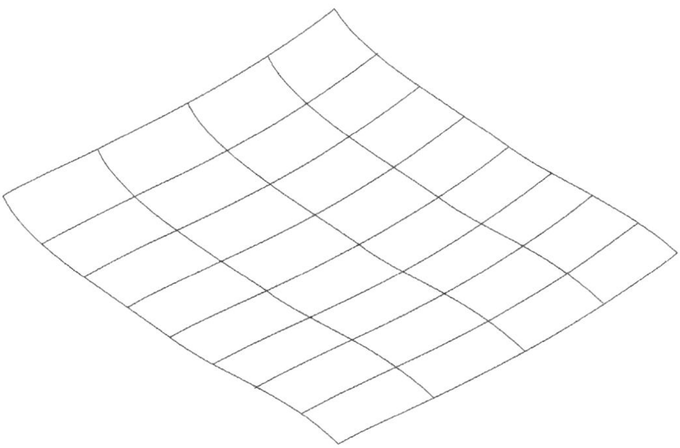

FIGURE 10.19 Two surfaces meeting with G1 continuity.

FIGURE 10.20 Zebra analysis of G1 surface join.

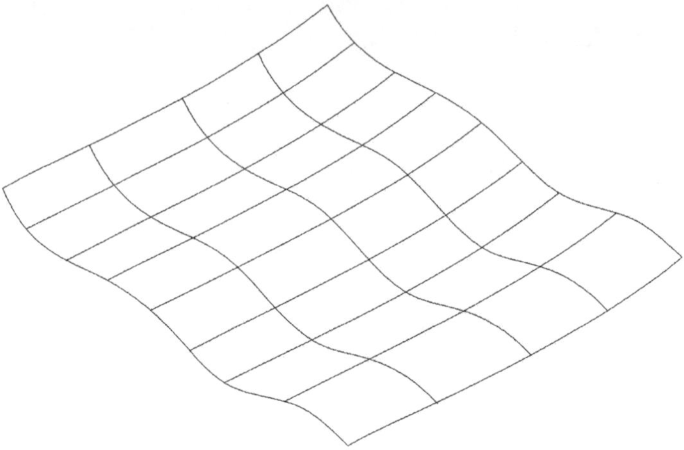

FIGURE 10.21 Two surfaces meet with G2 continuity.

FIGURE 10.22 Zebra analysis of G2 surface join.

stored internally with representations based on Bézier, B-spline or NURBS formulations and perhaps with a corresponding interpolation representation that will allow the use of different editing modes.

10.15.1 Resurfacing

Sometimes when many surfaces have been joined together it is useful to replace the multiple surface patches by one large surface that may be easier to manipulate and edit. CAD systems usually offer some facility for achieving this.

10.15.2 Blends

Using blends (frequently also called rounds because many are circular in sections) is a useful way of altering part geometry that allows designers to smooth general shapes. This is often done for aesthetic reasons or because of manufacturing constraints and it is especially important for parts that are to be moulded or cast, for example. Normally these are applied late in the design activity (especially if they are finishing operations) and perhaps grouped towards the end of the design where they can be conveniently edited or switched on and off. In practice it is often difficult for designers to fully understand the geometric implications of applying round and blend commands and it can end up becoming something of a black art or trial and error process. Understanding something of the surface geometry and how blends are applied does help; however, even though the process of applying these will still involve some trial and error and the exercise of some patience to get good results. The most common rounds applied to normal parts can be conveniently thought of as 'rolling ball' rounds. The effect is to 'fill' the volume between two faces and an imaginary ball is placed so as to touch them tangentially as shown in Figure 10.23.

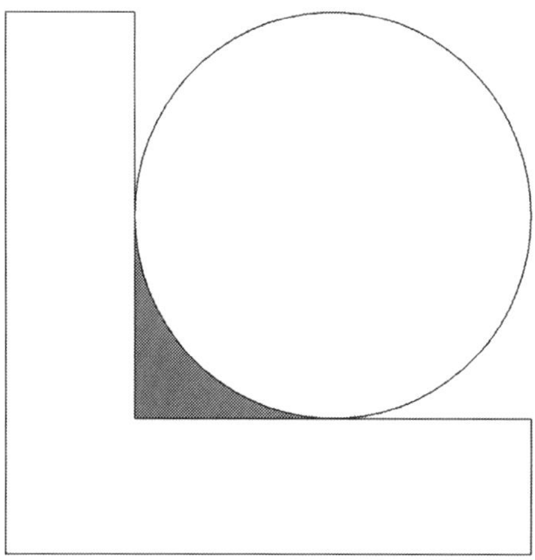

FIGURE 10.23 'Rolling ball' blend.

The shaded area shows the result of a simple round operation in 2D. Clearly the resulting blend will be tangential (G1) with respect to the parent faces. Where more complex faces are used we may also wish to specify curvature (G2) continuity.

Modern CAD systems are likely to offer a range of options to round or blend features, especially at corner positions where two rounds or blends may meet. Typical options include specifying whether the joint should be mitred (i.e. the 'rolling ball' will continue along the tangent of each edge) or not (where the 'rolling ball' will roll around the corner). The best way to learn to use rounds is simply to try them out on simple shapes and observe the results. Most of the time the results can be easily foreseen by imagining a ball rolling round the edges formed by touching faces on the outside of the part (for concave edges) and the inside of the part (for convex edges).

10.15.3 Curves Revisited

Now that we have considered several surface types we can revisit our means of constructing curves. In addition to the basic ways of constructing curves from scratch (sketching, keypoint, from tables of coordinates and so on), we can generate curves by considering interactions between any predefined geometry. For example, we can create complex curves by intersecting two surfaces or by projecting or wrapping existing curves onto surfaces and we can draw curves directly onto surfaces. We can generate curves that are based on surface metrics like curvature (e.g. isoclines).

10.16 COMMENTS

Working with curves and surfaces can be a challenging activity, mostly because it is a complex field mathematically and programmatically; however, understanding some of the mathematics behind the tools in CAD systems can help the user to work reasonably effectively. In addition to this there are some useful rules of thumb that can be adopted to help further the user's quest to design robust and tidy models. Here are a few.

- Use four-sided surfaces (or 'patches') where possible. Using patches like the ones described earlier (with two parameters u and v) usually makes for fairly robust models. These tend to be easier to attach other features to, or to simplify and add fillets and other downstream features to.
- If you are ultimately using a set of surfaces with which to create a solid, try to do this in one operation. In other words, do whatever work you need to create a good surface model and then thicken it or use whatever methods your CAD system has for turning construction surfaces into solids.
- Keep surfaces as simple as possible when creating them, possibly oversized, and then trim these to fit your model. You can create surfaces that cut through many existing faces of the model and then subsequently

use the intersecting curves to trim the new surface to size and to remove any excess geometry from the existing surfaces and solids.
- Use some surfacing tools with care. For example, attempting to extend a surface that is created with complex splines might result in self-intersecting surfaces or surfaces with very tight bends. This is because of the constraints on such commands that are imposed by continuity of curvature.

This chapter has given a very brief introduction to curves drawing and surfacing but it should be useful, particularly to newbies. To understand more you might have to consult entire books that specialize on the subject. It might also be helpful to consult Wikipedia and other web sources as these contain animations of the construction of some of the common splines discussed here, for example, Bézier, B-spline and NURB curves and surfaces.

11 3D CAD File Types and Transfers

11.1 INTRODUCTION

At some point in almost all designers' careers, they will come across the challenge of having to deal with CAD or physical models that have been generated previously, by someone else, but which need some or all of their information incorporated into a design. These may be files generated on the same CAD system by a colleague, for example, or they may be files that have been sent by a customer or supplier and may be in any form, including filetypes generated from scanned physical models, either for validation or for reverse engineering. The files may contain models that are developed in either imperial or metric units or both. Alternatively, a designer/modeller may be the source of CAD information that is to be retained over a period of time or sent to a customer/supplier.

Increasingly, CAD models are forming part of the contractual process that was previously the domain of engineering drawings. **ISO16792** gives guidance on procedures for augmenting drawings with CAD data. Most modern CAD systems have a number of tools that will allow the user to save or open files in a variety of formats but choosing the best one for a particular task can give rise to a bewildering array of options. Users reading others' CAD files will frequently ask themselves—and anyone else who'll listen—questions like the following. 'Why will the file not open?' 'Why is the model huge (or tiny)?' 'Why can't I edit this model?' Understanding something of the basic file types for part modelling can help users answer these questions.

The most obvious choice to the modeller when supplying models to others is probably whether to utilize 2D or 3D technology. **2D models** are traditionally associated with sheet drawings but many CAD systems allow user to model in 2D on large or near infinite sheets so that modelling can be done without the need to be constrained by **drawing sheet** sizes and **scale** limitations. **3D models** can support one, or several, or all of a variety of data types, **point clouds**, **meshes**, **voxels**, **faces**, **solids** and **meta-data**, for example. Let's consider some of the most common and emerging options out there.

11.2 BASIC CAD FILE TYPES

2D CAD models may be characterized as either structured or layered types. Structured models make use of the type of entity that is being used in a drawing, dimensions, GD&T/GPS boxes, callouts, cutaway views or detail views, for

DOI: 10.1201/9781003395072-11

example. Layered drawings store different aspects of drawings on different layers of a drawing and these are under the control of the designer who may, for example, put all of the dimensions on one layer and the outline geometry on another. The number of such layers is usually limited.

There are a large variety of 2D file formats that may be either unique to a vendor or in a standard format. One of the most common file formats is a vendor-specific one which has since been adopted almost universally as a standard, the **DXF** file that was developed in AutoCAD. This is a layer-based format. Most mid-range and high-end CAD systems will readily open and respect the layer data in DXF files.

3D CAD files, when created by a modeller from scratch, are usually based on **boundary representation**s (**B-Rep**) where a data structure is created that represents solids bound by faces that are bound by surfaces and edges at the simplest level. Higher-level file types may incorporate other special boundary types and will be able to store and interpret spline curves and surfaces, for example. Common commercial kernels are becoming increasingly sophisticated and may include attribute data such as common thread dimensions and tolerances. They may also hold information regarding assembly structures for products and point to other associated files.

Other types of 3D format may be based on point or mesh data that are obtained by either being generated from a B-Rep model or being read in from scanned (e.g. laser scanner or CMM) data. Point clouds are simply collections of many (thousands or millions) of coordinate points that represent a 3D shape. Mesh data types are made up of polygons—most commonly triangles—that are bound by their edges, and in turn, vertices. Mesh files may or may not be guaranteed to be watertight, the triangles may not stitch together perfectly and this is often as a result of rounding errors on the positions of vertices. While this type of file is 3D, it is not always regarded as 'solid', and while it might be useful for some spatial analyses tasks, or for rendering, it is not particularly useful for calculating solid based values of mass or moments of inertia for example, or for carrying accurate dimensions.

As with 2D files, the majority of common 3D file types in use have been developed and maintained by standards bodies or by vendors.

11.2.1 Neutral versus Vendor-Specific File Types

Many of the truly **neutral file formats** in common use are based on standards that have been developed by government or independent bodies, though often in conjunction with commercial third-party vendors or potential users. The commonly used, though now quite dated **International Graphics Exchange Specification** (**IGES**) format, was originally developed by The United States Air Force but published and maintained and made freely available by the National Bureau of Standards. The **Standard for the Exchange of Product model data** (**STEP**) format is based on the **ISO 10303:21** standard, although STEP's development happened around a history of academic/industrial research and development projects.

Neutral file formats have the advantage that they are generally freely available and not controlled by a particular company; however, if they are not backed by major commercial companies they can lack the traction that's needed to ensure widespread adoption.

Vendor-specific file formats may be in a simple form and governed by a set of rules that can be adopted by other users, or they may be encoded and require specialist software add-ins to create and manipulate the data structures, often requiring paid-for licences. Autodesk's development of *.dxf and *.dwg formats and their widespread use by other vendors led to these becoming de-facto standards. Often vendors see the widespread adoption of their CAD data models as a plus in the sense that it helps create user communities based around their products and they are happy to encourage the use of their in-house standard formats. Siemens' **JT** format, for example, is becoming a popular high-level product data format capable of storing a wide range of 3D geometry types as well as high-level meta-data on assembly structures, materials data and various other useful information making it a popular choice in the **PDM/PLM/digital twin/Internet of Things** technology areas. Similarly, vendors such as Dassault and Siemens also produce kernel 3D modellers (**ACIS** and **Parasolid**, respectively) that are in widespread use in very many CAD/CAM/CAE applications.

11.2.2 Text versus Binary

3D data formats are often offered as either text or binary (or both) file types. Text-based formats (e.g. STEP) are generally easy to read, and this can be important when trying to troubleshoot any problems that end users might have in opening translated CAD files. It is sometimes claimed that because of the way they typically store geometry; however, they are more likely to suffer from rounding errors and this can cause problems where the geometry is not watertight. The files produced in text format can also be rather large.

Most binary formats (e.g. **DWG**) result in smaller file sizes than their text equivalent but they cannot be readily opened and inspected with simple text editors.

Where CAD file types are offered in both text and binary format they will generally offer two types of file extensions that display which type they are for example, the *x_t and *.x_b format Parasolid save files.

11.3 TRANSLATORS

Many design projects will involve working with multiple CAD systems and fortunately most major systems will support this at some level although this may mean translating between file types. Most systems offer a wide range of tools to export and import data to and from other applications. The following paragraphs will give some support for users faced with CAD data transfer tasks. For specialist support, there are companies who have considerable expertise and who specialize in CAD data transfer, offering services such as web-based applications or who provide specialist installed software. They may also offer consultancy-based support for specific projects. This can be particularly useful for large batch operations, for example, building web-enabled catalogues of parts for collaboration or sales-based projects.

11.4 COMMON DATA TYPES FOR 3D PRODUCT INFORMATION

There is no one best format to use when transferring CAD models; however, most users tend to use the highest level option—the one that saves the most data—that is available. This is not always the most desirable practice, however. For example, if you have used a native CAD system to build a complex assembly with simulation information (FE, kinematic and so on), and it is needed by the receiver of a model then it is of course wise to use the native CAD format; however, models that require less sophisticated data may use simpler file formats.

Many of the common data types are actually saved or exported with more than one file and these should be kept together. It is common to save a geometry file and a second configuration or log file that contain the preferences used such as the measurement units used or material types.

Roughly, in order of descending sophistication, starting with the richest CAD specific tools, the following common data types are widely used.

11.4.1 CAD Vendor Specific

The major CAD vendors such as Siemens (Teamcenter, NX and Solid Edge), Dassault (CATIA and SolidWorks), PTC (Creo and Onshape) and AutoDesk (AutoCAD, Fusion, Inventor), all have their own file types and these may be broken down according to their application, commonly, assembly models, part models, sheet models, weld models and 2D or drafting models, to name a few. These formats typically contain rich information on product structures, product manufacturing information (PMI) and may carry many data types to convey simulation results and a range of geometric data such as B-Rep, mesh or voxel formats.

Some care still has to be taken even when transferring models between like-systems. Firstly these formats are generally not backward compatible so you may need to make sure that the receiving system is at least at the same version level as the originating system. Secondly, it is important to copy all the files that the originating system used, for example, the configuration or log files, often saved as *.cfg, *.log or *.txt files.

11.4.2 Standard Product-Level Formats

Some of these systems were originally developed simply to allow the transfer of geometry from one CAD system to another; however, with the increasing demand for product structure information to deal with assemblies and information relevant to product data management (PDM and PLM for example) systems and the desire to develop digital twin models, the need for rich common data types has led to formats developed from industry as well as standards bodies. Examples follow.

11.4.2.1 JT

Developed by Siemens, this format is extremely popular for two reasons. It is a very useful and flexible format for carrying geometric and meta-level product information, and it is used widely in PLM systems to store CAD data, especially when the analysis of large datasets made up of many CAD models is required. JT files can contain 3D information in a number of different formats, natively as Parasolid (see 'Kernels' below) files but also as **3D PDF** files or those from other vendors as well as various mesh and other formats. It has full support for NURBS. Probably the most commonly used PLM system in the industry is based around the Teamcenter platform and JT files work well with this. Typical analysis of large datasets made up of CAD files includes tasks such as shape-based searching for models or automatically predicting component clashes due to design changes.

As well as being a very useful format for 3D geometry transfer, it carries support for **model-based definition** (**MBD**), allowing integrated PMI, and enabling **model-based systems engineering** (**MBSE**) as well as the concept of the **model-based enterprise** (**MBE**). The format is therefore positioned to enable the realization of a digital twin representation of products that can support modelling of products from design through manufacture, validation and disposal/recycling.

JT is a neutral binary file type; it is not based on a specific CAD system and can support data from multiple CAD types. It is efficient in that it can save files with minimal need for space.

JT documents are saved as *.jt files.

The richness of the JT format means that there are normally lots of optional ways to store and transfer product models in this format and the user should choose the best set to ensure that the information needed by the target systems is included. Options may include, for example, the units of measurement used, the JT version number, whether all the assembly/part files should be saved as individual or separate documents, whether sheet metal files should be as-designed or flattened, whether PMI data should be included, whether the precise geometry is required, whether simplified part/assembly models are required. The JT files generally keep any optional header data within the files themselves so there is no need for additional log files, for example.

11.4.2.2 Quality Information Framework

One of the most modern formats, Quality Information Framework (QIF) was, as the name suggests aimed largely at metrology and downstream applications in the product lifecycle, making it friendly for MBD applications. It supports PMI information, and it has an XML-based framework, giving it potential for integration with web-based applications. Although it is an interesting format, it is not generally available as an export option in many popular CAD systems to date.

QIF documents are saved as *.qif files.

11.4.2.3 STEP

The STEP standard was originally developed by researchers and industrialists in conjunction with ISO and was created as a successor to the IGES (see the next section) neutral file format. The complexity of this task has led to different forms of STEP, each aimed primarily at different industrial environments. For example, STEP AP 203 was the initial format and this supported assembly and part geometry and topology as well as basic configuration management. It was supported by the aerospace and defence industries. Later came STEP AP214, which is a superset of 203 and added support for colour, layers, textual annotation, construction history and tolerance data. The development of this application was driven largely by automotive companies.

STEP AP 242 is currently the latest incarnation and adds improved support for various forms of meta-data. For example, it supports PMI as well as 3D geometric constraints, kinematic relationships in assemblies and support for electrical harness design and 3D piping. A major aspect of 242 is its support for some lifecycle data. It is aimed at the adoption of MBD/MBE/MSE techniques for representing products where the computer model can be used as the major product definition. The widespread use of these functions continues to develop.

Step files are saved as *.step or *.stp files along with a *.log file (optional)

Optional log files can store information such as whether construction elements (bodies, sheets, wires points) are needed or whether PMI data is required (AP 242 only) as well as the units of measurement used and other additional user information.

11.4.2.4 Initial Graphics Exchange Specification

The Initial Graphics Exchange Specification (IGES) began development around 1979 under the auspices of the US Air Force and was subsequently published as a neutral format by the US National Bureau of Standards.

Now rather old, last updated in 1996, the IGES format suffers from the fact that its B-Rep structure is such that the faces that should form solid boundaries are prone to not being connected and this leads to non-watertight models. Likewise, the mesh structures it supports are also prone to disconnectedness. Although there are now IGES versions that better support solids, these are seldom implemented in the major CAD systems. The standard also has little or no support for PMI data.

IGES does support assembly structures and although it is old-fashioned from a technical viewpoint, due to its 3D geometry representations, it is still a very popular method of 3D file storage, and there is still a great deal of legacy data based on this standard, and so it remains important.

IGES files are saved as *.iges, *.igs or *.ige files and may have additional *.log files

Like STEP files, when saving IGES files, it is possible to also generate a log file that will store any options, for example, the units of measurement used, whether construction elements (bodies, sheets, wires points) are needed, how solids should be saved (analytics, NURBs or wireframe) as well as additional user information such as the user name, comments or the name of the originating CAD system.

11.4.3 Kernel Systems

Inside CAD systems there is generally a geometric engine or 'kernel' that takes care of creating, storing and saving the geometric data related to the CAD system. This generally doesn't carry substantial meta-data and is often supplied to a CAD vendor from a specialist creation company. Many, if not most, of the kernel systems in use today have their roots in university or commercial research groups. Some CAD systems have their own specific kernels, for example, Inventor, CATIA and Creo. Although there are many kernel systems, for file transfer purposes between commercial 3D solids modelling applications in the mechanical/manufacturing sectors, Parasolid and ACIS tend to dominate.

11.4.3.1 ACIS

ACIS was originally developed in the 1980s by a small UK company called Threespace Ltd and subsequently acquired by Spatial Corporation, now part of Dassault Systèmes, which developed it further. The ACIS modeller was designed from the outset so that it might form the geometric core of higher-level systems such as CAD/CAE packages.

The ACIS modeller has good geometric capabilities and was designed as a powerful B-Rep application from the outset. It can operate on and store B-Rep entities such as solids, sheets, wireframes, multiple bodies and non-manifold bodies. The output from the modeller results in watertight bodies. The modeller, at time of writing, does not provide substantial support for assembly hierarchies or for rich meta-data as it is normally embedded in larger CAD packages that deal with these aspects.

ACIS documents are stored as *.sat (text) or *.sab (binary) files.

Although ACIS is ideal for transferring geometric models to other ACIS based applications, many non-ACIS packages can read/write import/export ACIS files.

11.4.3.2 Parasolid

Like ACIS, Parasolid was developed in the UK in the 1980s, and both of these kernels can have their roots traced back to the earlier Romulus modeller, the first-ever commercial kernel, which in turn can have its roots traced back to the University of Cambridge. Not surprisingly, Parasolid and ACIS have some similarities.

Parasolid has excellent B-Rep modelling capabilities at its core and can operate on and save NURBS geometry. It supports assemblies, sub-assemblies, component models (as design bodies) as well as construction geometry, and it has some support for meta-data.

Parasolid documents are stored as *.x_t (text) or *.x_b (binary) files.

Parasolid is the geometric kernel used in many of the top 3D CAD systems including NX, SolidEdge, SolidWorks, Onshape and many other applications.

11.4.3.3 3D PDF

Not a kernel but 3D PDF files are useful for downstream viewing and mark up of 3D models for parts and assemblies and at that level they are useful collaborative tools. The format originally used the internal 3D storage format based on *.u3d files. The modern versions however use the *.prc format for the internal storage and manipulation of various types of 3D data including B-Rep and mesh formats. The format does require the use, though freely available, of Adobe's Acrobat Reader but the 3D PDF formatted models cannot be opened in most common CAD systems, so editing and manipulating the geometry is not generally a common or realistic option.

3D PDF files, like ordinary 2D PDF files are saved as *.pdf documents.

11.4.3.4 DWG/DXF

Not a kernel as such, and predominantly 2D, DWG files can store some 3D data, and there is a vast number of these files out there. They are not widely offered as save file types in top-end CAD systems other than AutoCAD-related applications, but they are readable by many CAD packages, although the user may be required to do some repair and rebuild work on imported files. Although DXF files are 2D, these are supported by many, if not most, CAD packages and they are also ubiquitous. Many CAD packages have tools to help the user build 3D models from DXF files.

DWG/DXF are stored as *.dwg or *.dxf files although DWG files might also be encountered as *.bak or *.dws files.

11.4.4 Mesh

There are many mesh formats, and they differ in terms of whether they are fully closed (watertight) and the typical facet aspect ratios that they generate and allow. Nearly all the common mesh types in use for file transfer are based on triangular facetted meshes.

11.4.4.1 STL

STL is probably the best-known triangular mesh used for representing 3D shapes. It is a very flexible format, but the quality of files that can be generated in this format can vary significantly. STL files were originally developed for stereolithography applications, and they are still widely used for additive manufacturing work.

STL files are very simple in structure. They are made up of individual triangle entities whose three vertices are defined by Cartesian coordinates. Using the right-hand rule for the triangles, the ordered vertices define a normal unit vector (see Figure 11.1) that points away from the surface into free space. This means they can define solids. STL file generators make the triangular meshes according to a few other rules too. The vertices are taken from the surface of a solid shape or sampled from single data points. Thus, the space defined by the gaps between connected vertices is undefined but approximated with straight lines. The vertices are generated so that each triangle must share two vertices with its neighbouring triangles. This helps to keep useful aspect ratios (but doesn't guarantee it) of the triangles (although they are not equilateral). This information is not written out to the files themselves, however, so each triangle is an individual entity and care

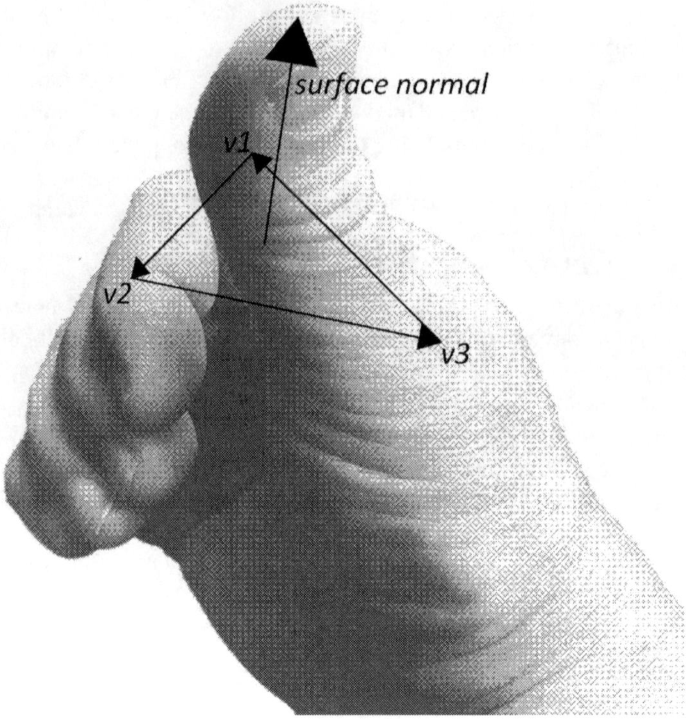

FIGURE 11.1 Right-hand rule.

must be taken when importing files that adjacency tolerances will recognize connected triangles. Even then there is no guarantee that an STL file will be closed or watertight.

Many CAD systems will offer the option of saving STL files, and these are useful as shape approximations for visualization and may be used by some numerical control applications. Better approximations to the nominal geometry can be achieved using smaller triangle sizes; however, this can result in excessively large and unwieldy files.

Many 3D scanning packages will offer support for building STL files from scanned point clouds, though these often result in disconnected surfaces and require some skill by the user to convert into meaningful closed solid object models, for example, if they are to be 3D printed.

They can be generated in both text (ascii) and binary formats, both as *.stl files.

11.4.4.2 OBJ

More modern than STL files, though not yet as popular as STL, are OBJ files. This powerful format makes possible the ability to store good quality additive manufacturing-ready files without excessive file sizes. However, this very much depends on the software used to create the OBJ file.

OBJ format can, in principle, store geometry as a tessellated surface using polygonal face elements such as triangles or quadrilaterals. However, they can also represent curved shapes using free-form curves and surfaces using B-splines such as NURBS (see Chapter 10) but support for such constructs is not readily available in most third-party CAD implementations in the author's experience. Although the ability to read/write OBJ files is available in most CAD systems, some of the options for file storage are not always implemented. In most cases, however, the files can store construction geometry as well as material and colour information and the generating software may offer options for trade-offs between precision and file size.

The OBJ format saves files as *.obj files along with additional *.mtl and *.log files where required.

11.5 OTHER FORMATS

There are very many 3D file formats. The examples given previously are those most commonly used in the mechanical and associated industries such as industrial equipment, aerospace and automotive work. In other industries such as architecture and civil engineering there are a number of other formats that are used and we have not covered them here. In the past, the construction industry tended to be based around 2D formats, with surface modellers for 3D depictions; however, richer models are now commonplace with full 3D solids models and meta-level information commonly being part of **Building Information Management** (**BIM**) systems. So, technologies such as 3D solids modelling, for example, are becoming more commonplace in construction and in shipbuilding.

11.6 GOOD PRACTICE FOR FILE SAVING/EXPORTING

Knowing the best way to save a file depends very much on the exact circumstances and reasons for the transfer; however, most CAD engineers will have a few good practice rules that they've found to help minimize problems downstream, and here are those of the author.

- It all starts with having good models in the first place. Using sound geometric relationships and well-structured models will help minimize overly complex geometry, especially on curves and surfaces.
- Garner as much information as possible about the target system. There is no right format to use as it all depends on what and how the model(s) will be used. Do you need meta-level information? Do you need to transfer construction geometry? Is there other information in your model that's not needed in the transfer file, for example, generative design shapes or FE meshes and analyses?
- If you are going to use native CAD files or MBD level (STEP, JT) ask yourself what versions of software should you save in? Sometimes it pays

to save in a past older version of a system or component application so that it will be more readily accessible to target systems. This applies to the CAD system and any component files, for example, Parasolid, ACIS, STEP and so on.

- Make sure you communicate as much as possible about the details of any save, for example, any defaults or options presented to you during the save/export. Include *.cfg, *.log, *.mtl or other files or make notes and include your own text notes with the saved file(s).
- Save as much as you might need but no more if you want to keep file sizes as small as possible.

11.7 COMMON PROBLEMS OPENING SAVED/EXPORTED FILES

Unfortunately, there is no simple 'one technique cures all' solution to opening files containing 3D geometry. The most difficult scenario is being presented with a file that is in some seemingly known format but which you have little information about its provenance. You may have no information about the source system's version and no additional *.log files, for example. Experience will probably teach you some useful techniques for dealing with common problems in one's own environment, but some of the problems and solutions the author has found relatively often are given here:

- **A file will simply not open/import**. At some stage you might simply have to give up but before you do there are a few things to try. Obviously if you can communicate with the source agent you can try again with different formats/options/versions. You might try and explore if there are import options you can adjust and if this doesn't work you could try to open the file with a different target system, which may mean downloading a trial copy of a system that you can then evaluate to see if it suits your needs.
- **The file opens but there are faults with the geometry, for example, missing areas of the part/assembly.** These are usually also caused by mismatches between the source and target systems and the way they build geometry. This can be a common problem when transferring 3D curve/surfaces. Although source and target systems might both use NURBS, for example, the way that a surface is built might differ, using different knot vectors and continuity options. In these situations, the target CAD system might have some repair/rebuild options that can be tried, or you may have to make some surface repair patches of your own to obtain a legitimate file.
- **The file opens but the sizes on the part/assembly are wrong**. The most common cause of this results in sizes being scaled wrongly either by a factor of 1000 or by 25.4. This is normally caused by dimension options in the source file being different from the target and is often because *'log

files have not been transferred with the geometry file. In the case of errors of a factor of 1000, it's because mm to metres or vice versa have occurred. In the case of files being out by a factor of 25.4, it is because a decimal/ imperial units mismatch has occurred. These can usually be fixed by changing the options on open/import in the target CAD system. On first encountering these problems it is sometimes difficult to spot the type of scale error, but it usually only comes down to the two cases mentioned and therefore they can be easily tested for.

- **The file opens but the part doesn't appear or cannot be edited**. This is often the result of importing the wrong type of data, even though it may be in a commonly shared file type. For example, mesh representations have been imported but not B-Rep data using a file type that supports both.

11.8 COMMENTS

Although modern CAD packages normally carry a host of tools for transferring 3D data between systems, the activity of sharing data can be a deeply frustrating experience. Taking time to try to clarify formats at the earliest opportunity will usually pay large dividends down the line. High-level data structures for product-related data are constantly in a state of flux and subject to frequent tweaks and upgrades as well as being open to competition from new formats.

As well as sharing models, it is sometimes required that actual physical products or artefacts are presented that need to be turned into models. This process is useful for reverse engineering and for validating parts as described in **ISO 1101**, and it is the subject of Chapter13.

12 Assemblies and Standard Parts

12.1 INTRODUCTION

Assembly drawings and models usually differ from simple part drawings and models because they depict information about more than a set of individual parts. They model relationships between parts, and so there is usually less emphasis on dimensions and tolerances and more on **product structure**. Assemblies can become very large and unwieldy in CAD systems, and this too affects how assembly documents are drawn up. Fortunately modern CAD systems have tools that help designers dealing with both large assemblies and with the unique demands of displaying assembly related information. It's worth a short chapter to consider some of these, although it's important to note that CAD systems differ markedly in their level of support for assembly modelling.

Assembly drawings are more likely to benefit from some of the advantages of using 3D models. Techniques for creating **cutaway** and **sectioned** views are useful to show the insides of assembly structures. Figure 12.1 shows a common use of 3D exploded views to illustrate (imply) the structure of a simple pulley mechanism. Note that views like this, while useful to show the viewer how an assembly might be put together, provide little meta-data. They do not give details that might be shown in parts lists or Bills of Materials nor do they depict general assembly attributes such as mass.

Figure 12.2 shows a sectioned view of a pulley assembly. As is standard for such assemblies, part webs, nuts, washers and shaft components are shown without sectioning. Note that the parts list contains the paint item although this is invisible. In many CAD systems the easiest way to do this is to make a part but add no solids to it.

12.2 ASSEMBLY DIMENSIONS

The main dimensions and tolerances that are normally shown on an assembly drawing are those that relate to how the parts are assembled, for example, if the distance or angle between two parts has to be specified and measured then it should be shown on the assembly drawing.

12.3 FLEXIBLE PARTS

Normally in single-part drawings, the default would be to show a flexible part in its relaxed or unstressed condition. In assembly documents, it is frequently the case that parts are shown in some constrained condition. A spring, for example, would

DOI: 10.1201/9781003395072-12

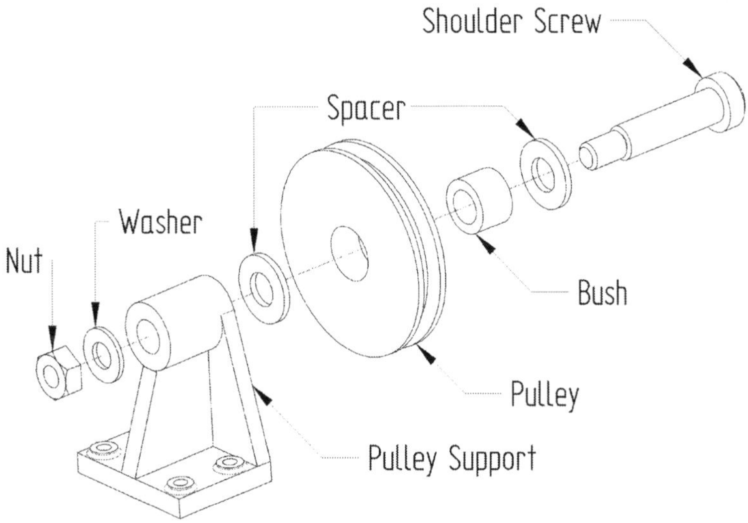

FIGURE 12.1 Annotated and exploded view of pulley assembly.

File Name	Quantity
PulleyHousing	1
ShoulderScrew	1
Spacer	2
Bush	1
Washer	1
Nut	1
Pulley	1
Paint	1

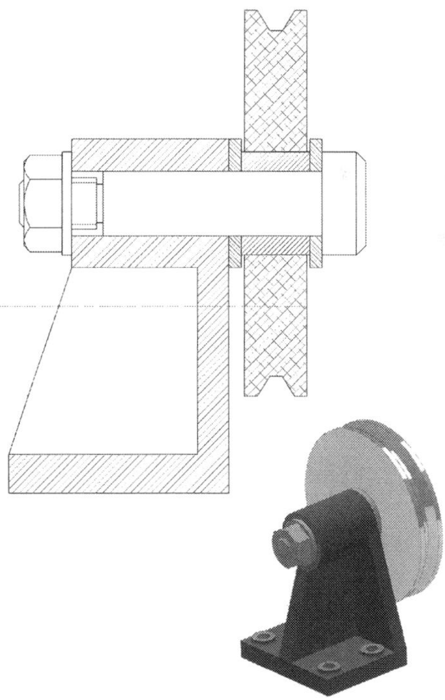

FIGURE 12.2 Sectioned view of the pulley with parts list.

normally be shown unstressed on a part drawing but may well be compressed or stretched to fit into a normal assembly state.

12.4 FAMILIES OF ASSEMBLIES

Sometimes on a drawing it is useful to represent an entire class, or family, of assemblies on a single sheet. For example, suppose we wanted to offer the previous pulley assembly as four different products each with a unique diameter of pulley wheel. We can do this on a single drawing sheet as shown in Figure 12.3.

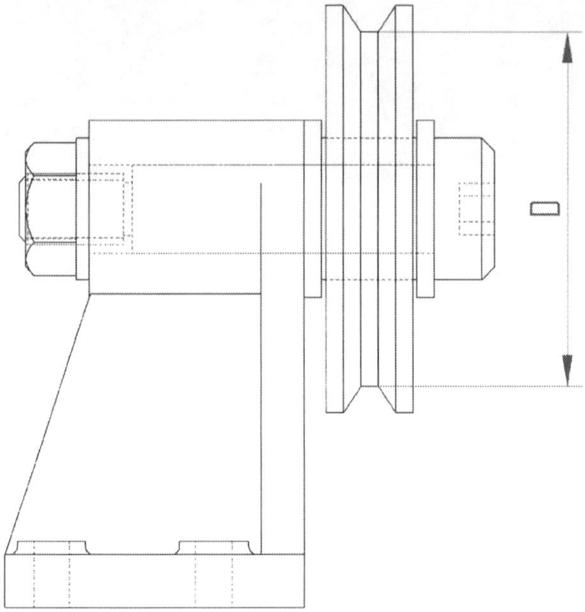

Part no.	Pulley dia D (mm)
8639/A	60
8639/B	50
8639/C	40
8639/D	30

FIGURE 12.3 Family of assemblies.

12.5 STANDARD PARTS

A definition of a standard part might simply be a part or assembly that is used regularly and which may be manufactured in-house or bought-in. Obvious examples include fasteners such as nuts and bolts or assemblies of fastener systems. Companies have a number of ways of saving time and effort with regard to such parts; for example, it may well be cheaper to buy such components.

When modelling with standard parts, there are further effort-saving strategies that can be adopted. Firstly, a library of standard parts can be developed and maintained in-house. This might include developing parametrized models, which can be scaled to required sizes as is common with fastener assemblies. Despite being useful, this does have the problem that someone will have to maintain such a library, and dealing with components that do not scale linearly means that either some non-trivial model building is required to create usable templates, or alternatively that explicit single models of each used size will need to be created. An example might be a series of M-size screws. As the thread dimensions are altered, the head, Allen key and other sizes do not scale proportionally.

A common second method of dealing with standard parts at the design stage is the use of bought-in components and their models. Many, if not most, suppliers of standard parts supply CAD models, which can be incorporated into the designer's model. These are usually offered in a variety of file types. See Chapter 11 for a description of file types that might be used.

13 Metrology, Associative Features and Reverse Engineering

13.1 INTRODUCTION

The study of measurement, or **metrology**, and the activity known as **reverse engineering** are inter-related and worthy of some overlapping joint treatment. In both cases the user is often presented with a physical part and asked to characterize it for the purposes of **validation** (metrology) or to build a virtual model that might be used for any number of purposes. In Chapter 5 the concept of an **associative feature** was presented, and this can be considered as a type of reverse engineered feature. Whether for metrology or for the digitization of legacy, competitor or other physical artefacts, the part(s) being studied will need to be measured in some way, either by **point-to-point** measures or by **scanning**.

Figure 13.1 shows three components that might need measuring so that digital models can be made from them. The garden gnome, Figure 13.1(a), would have been nearly impossible to create a convincing digital reproduction from using rulers, callipers or other length-measuring devices. The model used to create the image was, in fact, made by Dr James Young [1] using laser and white light scanners to create a triangulated mesh model with surface rendering. The second model, Figure 13.1(b), came from the archery world, in the form of a bow riser. Some aspects of this might have been conducive to scanning, for example, the general form, but it would have been a difficult task given that it has some sharp corners and narrow deep slots. It also has several precision features. Modelling this part required many measurements being made with callipers, depth gauges, micrometers and the like. It also required research and reference to data sheets to match measurements with known sizes for threads and other fittings. Lastly, the picture of the cube, Figure 13.1(c), is reproduced as an example where the simple apparent shape alone wouldn't dictate the nature of a model. If the part were a dice from a board game, point-to-point measures would probably suffice along with some dots and rounded edges and corners. If, however, the part is an engineer's slip gauge that needs documented and calibrated, it will require very demanding multiple measurements of length and shape and the calculation of parameters such as surface profile (see Chapter 16). If it is to be modelled for manufacturing it would require the addition of many GPS tolerances to control size and form.

DOI: 10.1201/9781003395072-13

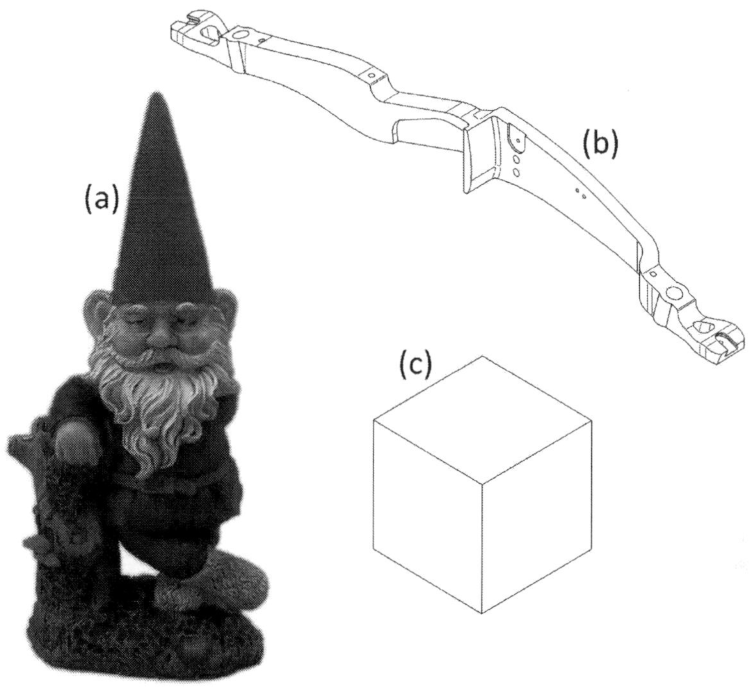

FIGURE 13.1 Example parts that might be measured or reverse engineered.

So, before embarking on a measurement task and deciding what equipment is necessary, we need to recognize that this must be done in the context of the purpose of the part and why we need a model.

13.2 METROLOGY AND GPS

Our interest in measurement in the context of the ISO GPS system is because, at some point, parts that have been designed with toleranced features will need to be validated once they have been manufactured. When batch sizes are large, then validation might involve making inferences about the population of manufactured parts from samples via **Statistical Quality Control**, or it might involve 100% inspection of parts. Either way, a manufactured part will need to be compared to the design to ascertain if it meets its requirements. **Go-NoGo** gauges are commonly used for in-process measurement, and, for large batch work, especially where 100% inspection is needed, but for lower batch sizes and high-value parts it is likely that some form of explicit measurement will take place.

The ISO standards (e.g. **ISO 5456**) use many adjectives to describe 'features' on parts, and this can be quite bewildering at first; however, the concepts encompassed by these descriptive terms are important. Perhaps one of the most fundamental descriptions of a feature on a part is the differentiation between **integral** and **derived**

features. Integral features are the features that make up the surfaces of a part, while derived features are imaginary features that need to be simulated, or derived, in some way. These include centre points, centrelines or the mid-faces of slots.

13.3 DIRECT MEASUREMENTS

Direct measurements include those made with traditional tools such as micrometers or vernier callipers to measure distances. Such measurements are taken with devices that are in contact with two faces on a part.

13.4 INDIRECT MEASUREMENTS—SIMULATORS

Measurements might also be taken indirectly, whereby a contacting surface is used to simulate an actual surface on the part, say a datum face where the part may be placed on a measurement table, datum face down and height measurements taken from the table's flat surface to features on the part. In this case, the table surface acts as a **simulated datum** and represents an extracted integral physical feature. This is a practice that can lead to errors since the 'flat' surface cannot be perfectly flat. It is common practice, however, to assume that the errors on the flat surface are small in comparison to those on the part and are therefore ignored or that an allowance is made for the table's flat surface variation in the specification of the original tolerance.

13.5 DERIVED FEATURES

Some measurements have to be taken to or from derived features such as centrelines or centre planes. In these cases, special constructs might be used, whose physical surfaces can be used to simulate the derived feature. As an example, a sphere can be fitted into a cone whereby measurements taken on the sphere can be used to establish a centre point that is in turn used to locate a point on the centreline of the cone. This might act as a datum feature. Other examples might include the use of a three-jaw chuck's rotational axis being used as an estimate of the centreline of the part being held by the chuck.

13.6 SCANNING-BASED APPROACHES

By 'scanning', we mean the practice of taking a large number of measurements and then building a model of the part. As an example, we might use **photogrammetry**, simple **laser scanners** or **touch probe sensors** placed on **measurement arms** or **CMMs** to establish the coordinates of a large number of points on the surfaces of a part. This data can then be used to establish estimates of features that might be used as datums or surfaces that need to be validated. In **ISO 5459:2011** parlance we call the original features as modelled in the CAD system nominal features (see Chapter 5). Nominal features can be integral features or derived features.

When a part is actually physically made, its instance will have what the ISO standards call **real integral features**, and it is these that we base the scanning or point measuring process on. The set of data points that have been collected form what is called the **extraction** of the part. This extraction can then be used to build surfaces in a variety of ways, none of which are trivial. For example, if an STL model is to be generated we must consider the data point density in comparison to the size of triangles that should be used to characterize features to avoid triangulation errors at corners and very close surfaces. If the only purpose of the extraction is to enable validation of the part then we can fit idealized surfaces to the extraction (planes, cylinders and so on) using the techniques described in Chapter 5. These then, are referred to as **associative** or **associated** integral features. We can also use these features to derive centre points, centrelines and centre planes that we can call 'a derived feature of the associated integral feature'.

There are very many different types of scanning techniques for dimensional and surface measurement, and many of these are subject to individual ISO standards, however; for now we generalize these as one group. Figure 13.2 shows an idealized workflow for how a part might be scanned for inspection or reverse engineering.

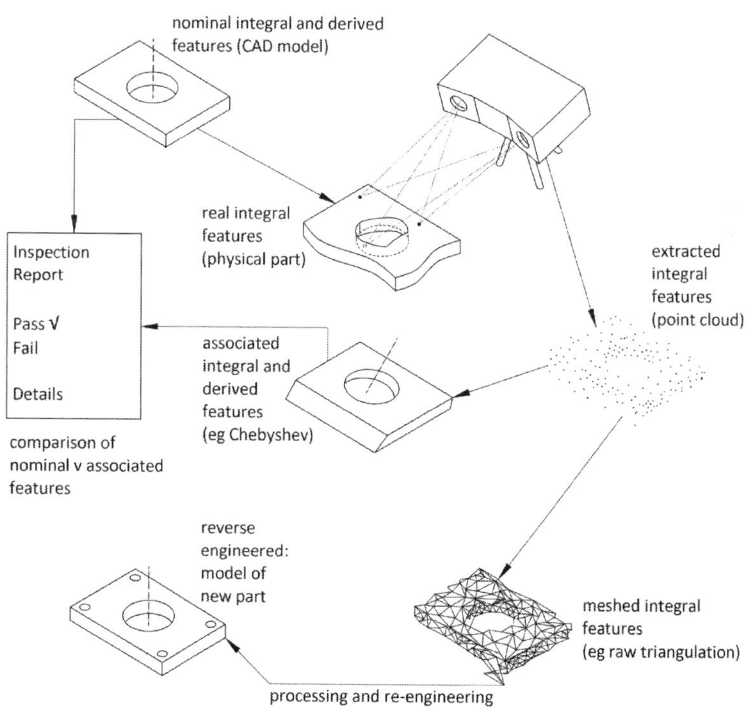

FIGURE 13.2 Scanning for metrology and/or reverse engineering.

13.7 MESHES

Most meshing activities using scanned data involve building triangular meshes on surfaces or tetrahedra in solids. These are, by necessity, irregular and hence largely unstructured. When working from point clouds, rough meshes can be generated reasonably easily; however, these methods vary considerably in the value of the final result. Several factors need to be considered as follows.

- A scanned point cloud is often depicted as a random set of points; however, these points are read from the surfaces of a part. Normally the first task in building a mesh is to group the points into individual surfaces. This information may already be available with the points previously grouped (e.g. as separate scans in separate files), or they may require some judgement by the engineer to avoid problems at edges and corners.
- The density of the point cloud may have a significant effect on how much detail can be associated with features. Small features might be missed by the scan. As scanning and computing technologies progress, greater point densities are made possible, reducing resolution-related errors; however, dense meshes are often undesirable because they result in large file sizes and generally contribute to the complexity of the engineer's task of building features from the data.

The lesson here is that triangulated or facetted models cannot be created automatically because there may be trade-offs between desired precision and feature sizes. Some CAD systems and packages (e.g. GeoMagic) help users to build models from point clouds, but this can still be a difficult and time-consuming activity.

13.7.1 TRIANGULATIONS

The nature of the triangles produced when turning point data into meshes can have a major effect on the usefulness of the mesh. Slivers, or long thin triangles which have very high aspect ratios, are problematic and may give rise to rounding errors in applications using the resulting files. A common and useful triangulation method that helps avoid such problems is called **Delaunay triangulation**. This uses an algorithm that maximizes the minimum angle of all the angles of the triangles in the mesh. Figure 13.3 shows how this works. A set of four points, P1, P2, P3 and P4, is evident. A Delaunay triangulation is one that meets the criteria that in the triangulation any point should not be inside any of a set of circles that are made by circumscribing each triangle, in other words, circles defined by three points (vertices). In the figure we can see two triangles; T1 has points P1, P2, P4 and is circumscribed by circle C1 while T2 has points P2, P3 and P4 and is circumscribed by circle C2. This is a valid Delaunay triangulation since none of the points lie inside any of the circles.

A more complex Delaunay triangulation is presented in Figure 13.4. This method is useful because it results in fairly robust meshes that avoid narrow angles and therefore the 'slivers' mentioned earlier.

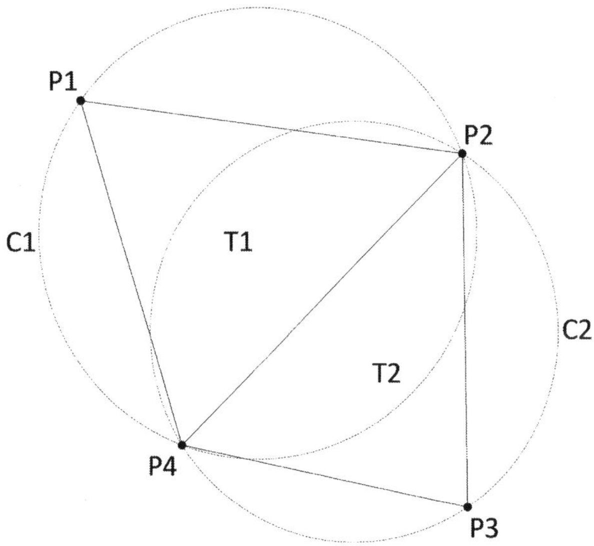

FIGURE 13.3 Simple Delaunay triangulation.

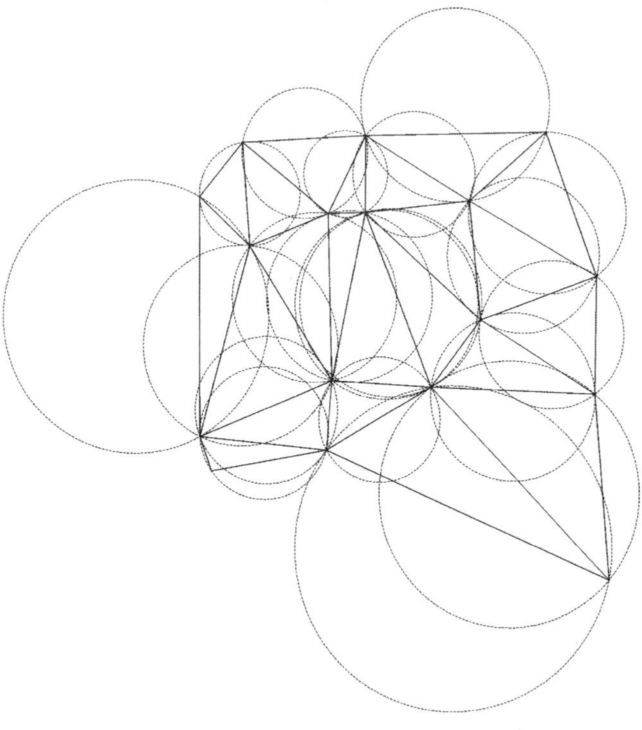

FIGURE 13.4 A more complex Delaunay triangulation.

A further important factor in creating meshes from point clouds is the desire to make models '**watertight**'. A watertight model is one where there are no gaps between the triangles, and it can therefore be used to simulate solid geometry. Watertight models are necessary if the part is to be 3D printed. Good meshing algorithms like the Delaunay triangulation help to make watertight models possible.

13.8 COMMENTS

In this chapter we have rushed through some important points in the process of reverse engineering. We have not described the myriad of scanning technologies that now exist but have drawn some general and important points that might help the reader better understand the processes involved and their inherent limitations. Note that the processes used in part scanning can require consideration of filtering techniques which may be dictated by the equipment being used. The **ISO 16610** series consists of a large number of individual related standards, the content of which is beyond the scope of this book.

13.9 REFERENCE

[1] Young, James; Undergraduate Meng Project, "Drone Scanning", Submitted to School of Engineering, University of Edinburgh, 2014.

14 NC Code

14.1 INTRODUCTION

Numerical control (NC) code is used to control nearly every type of manufacturing machine, whether cutting, adding or even moving material. From high-end \$1M+ machining centres to desktop 3D printers or routers, the control commands are very similar. In high-end machines there is usually a separate control unit attached, which may have considerable computing power, while desktop machines are often controlled more directly from a PC. High-end CAD packages usually offer some **computer-aided manufacture (CAM)** functionality or add-ins and additional packages might be used to help generate NC commands. The degree of computer centralization means that there are various terms for NC environments such as **CNC** for computer NC or **DNC** for distributed NC (computer network). Increasing use is also being made of **cloud-based NC code** preparation and management.

CAM packages offer considerable help for those tasked with developing code to control manufacturing machines. They offer the ability to model parts and stock material as well as tool and 3D machine models to enable complete simulations of machining processes. In doing so they encourage programs based on complex geometries that are efficient and safe, that is, those that can avoid tool/part/machine collisions. No matter which format is used to build high-level simulations, the last step in the process is normally called **post-processing**. The post step turns any package instructions into the raw NC code that is needed to control the specific machine that the work will be realized on.

NC code consists of line numbered geometric codes (**G-codes**) along with various miscellaneous codes (**M-codes**) as well as other parameters and X,Y,Z and I,J,K coordinate data. A simple example will demonstrate how NC codes are used.

14.2 EXAMPLE NC PROGRAM

Figure 14.1 shows a very simple part that is to be cut from a block of 40 × 40 × 20 mm.

The following code might be used to cut the step feature at the front of the part.

```
%                           /data start/
* 1001.NC                   /comment: program number and filename/
* T1 D=12—flat end mill     /comment: tool details/
N10 G17 G90                 /xy plane, absolute coordinates/
N11 G71                     /mm units/
N12 T1 L25 D12 M6           /change tool, select tool#1/
```

DOI: 10.1201/9781003395072-14

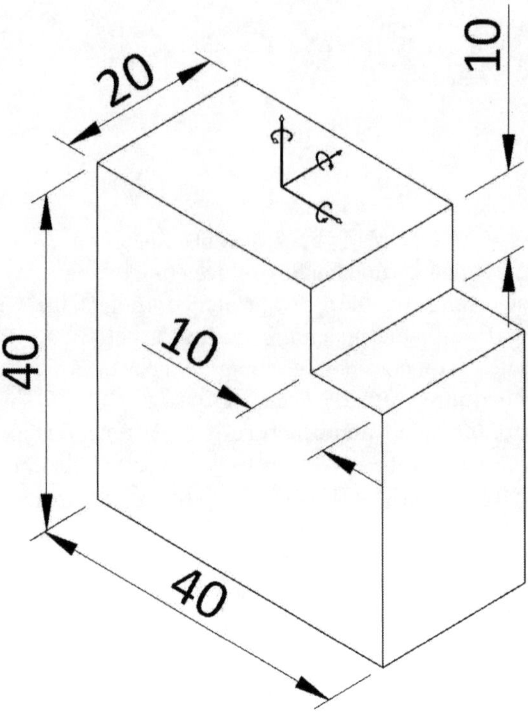

FIGURE 14.1 Example part to be cut from 40 × 40 × 20 stock material.

```
N13 S5000 M3        /spindle on CW, speed 5000rpm/
N14 M8              /flood coolant on/
N15 G0 Z5           /rapid move to Z5/
N16 X16 Y-17        /rapid move to X16 Y-17/
N17 Z-10            /rapid move to Z-10/
N18 G1 Y17 F330     /move to Y-17 at federate 330 mm/min/
N19 G0 Z15          /rapid move to Z15/
N20 X0 Y0           /rapid move to X0 Y0/
N21 M5              /spindle off/
N22 M9              /coolant off/
N23 M2              /program end/
%                   /data end/
```

14.3 FILE TYPES

NC programs are often stored with a variety of filename endings, **.nc** being the most common. Variants include. tap for use in PCB manufacture and. gcode, which is commonly used in additive manufacture. Most files refer to simple text-based formats; however, increasingly, NC1 file types are being used, and these are binary encoded files which can be read by proprietary machine tool controllers and these are capable of carrying additional specific machine and fixture information.

14.4 COMMENT

NC code is widespread in the manufacturing community and it is therefore important for any engineer to be aware of it, even if they never have to write a line of such code. If parts are to be made on any sort of modern machine, there's a very high chance that someone will have to develop a program. Their life will be made easier if the designer knows how these will be developed and makes their drawings or models NC friendly. This can be done mainly using a suitable dimensioning style, for example, by considering whether whether absolute or relative coordinate-based commands might be used. Although NC codes are considered 'standard', there is plenty of scope within these for application- and machine-specific codes to be used, so if you have to develop a program it'll be necessary to refer to a reference manual for the machine that'll do the cutting, depositing or positioning.

15 Additive Manufacture (AM)

15.1 INTRODUCTION

Additive manufacturing (AM) is a term that covers several technologies but in general refers to processes which work by building up layers of 2D shapes to make 3D forms; hence, some of the methods are sometimes referred to as **3D printing**. The various technologies in use are being developed at pace, and they offer huge current and future potential benefits in many aspects of design and manufacture. Prototypes can be prepared quickly, hence, the term **rapid prototyping**, and functional parts can be made in shapes that have hitherto been difficult to characterize at the design stage or realize at the manufacturing stage. To reap the benefits of AM technologies, it is necessary to adopt some new ways of designing and conveying design information.

15.2 WHY USE AM?

There are many reasons that AM might be used; some of the most common are as follows.

15.2.1 Machine Type Parts

Part shapes that may have traditionally been made by machines such as milling machines might be prototyped using AM, or, increasingly, individual replacement parts may be fabricated, especially in areas where access to machine shops might be limited.

15.2.2 Improved Part Designs

AM allows for improved part designs with few restrictions on general shape, for example. New replacement parts can be designed with complex geometry, either from scratch or from part redesigns. These might be manually driven, for example, by removing material that is not seen as necessary, possibly as a result of FE analysis.

15.2.3 Generative Design

Most **generative design (GD)** methods work by adding block material to constrained part regions that will then be subject to simulated loading and subsequently removing volumes that are subject to low stresses. This is done iteratively

DOI: 10.1201/9781003395072-15

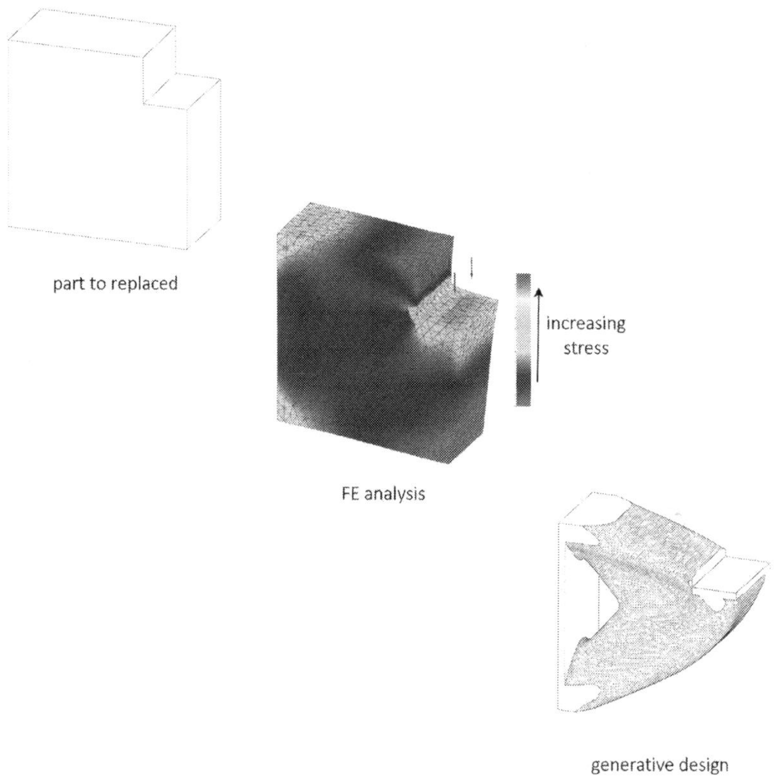

FIGURE 15.1 Generative design.

until some design goal such as mass reduction is reached. It frequently results in more organic shapes than might be expected using normal CAD tools. Once a new form has been created, the designer might use this as is or fit more well-defined shapes (e.g. spline surfaces) to it. GD shapes are often stored in file formats such as STL. In terms of integrity in most CAD systems they will produce good, closed STL files; however, GD formats are based on different shape representations and it is not always possible to easily generate an STL file from a GD. A generalized GD process is depicted in Figure 15.1.

15.3 ADDITIVE TECHNOLOGIES

There is a bewildering set of names that are used to describe AM technologies, and commonly several terms might be used to refer to the same process. The **ISO/ASTM 52900: 2021** standard seeks to clarify some of the terminology that should be used to adhere to ISO descriptions of the processes in common use. Currently, there are seven technologies that are recognized, although these vary greatly in terms of cost and volume of parts that they are used for, and there is some overlap in the methods individual manufacturers use. A brief overview of these and the various acronyms and terms used to describe them is given in the following subsections.

15.3.1 Binder Jetting

With **binder jetting** a substrate sheet of powder is laid down on a flat bed and then a jet nozzle deposits a liquid bonding agent onto the required sections creating a 2D solidified shape. The bed is then lowered, and a new layer of powder is laid down with the process repeating layer by layer. The process can be used on a wide range of material powders including metals, ceramics and polymers.

15.3.2 Directed Energy Deposition

Directed energy deposition (DED) involves a focussed energy beam being used to melt material, which is simultaneously deposited onto the melt locality as a powder or from a filament feed. With good control of the melt conditions, it is possible to attain good bonding between layers to make strong parts. However, process control can be difficult, and the parts can lack precision and hence uncertain surface finish and dimensional accuracy. It is common to use some form of post-process machining. DED is primarily used for metallic materials.

15.3.3 Material Extrusion

The most common and probably the best-known form of AM, **material extrusion**—aka **FFF, FDM** or **3DP**, consists of a material (usually polymer) fed from the filament (typically 0,4 mm thick) through a heated nozzle and deposited layer by layer onto a part. Material extrusion machines can be bought for a few hundred euros, but high-specification machines with improved resolution and process control can cost hundreds of thousands of euros.

15.3.4 Material Jetting

Material jetting involves microdroplets of polymer resin being delivered by a jet nozzle (rather like an ink-jet nozzle) and cured with, typically, UV light. It has the advantage that it can print in colour and is often used for model making. It is capable of producing very accurate and fine parts with layer resolution in the order of tens of microns (at the time of writing).

15.3.5 Powder Bed Fusion

Powder bed fusion (FBD)—aka SAF, SLM, EBM, SEBM, LBM, SLS, DMLS, ALM—uses laser or electron beams focussed on a substrate of powder which is melted, forming each 2D layer. It is possible to achieve good bonding between successive layers, and it is commonly used with metallic or polymer-based powders. In some processes a fluid is used to aid energy absorption, for example, with infrared energy sources.

15.3.6 Sheet Lamination

Originally developed for large-scale modelling, for example, architectural or geographic modelling, **sheet lamination** differs from the other AM methods in that it involves cutting out 2d shapes in sheet material and then using these to form layers to create 3D shapes, often using materials such as wood. However, more recently higher resolutions and strengths are achieved using paper/adhesive of metallic/weld-based layering.

15.3.7 Vat Polymerization

Vat polymerization, aka **SLA**, uses liquid polymer resins that can be solidified by light energy being used as the raw material. The liquid sits in a bath with a laser-based light source mapping out the required 2D layer shape on the surface that is to be solidified. The process is capable of making parts with high precision and excellent surface finish giving highly detailed parts.

15.4 DESIGN FOR AM (DFAM)

As always it pays to discuss requirements with the manufacturer at the earliest opportunity, ideally before or during the design process so that finished designs can be optimized for the manufacturing method to be used. Specific AM processes produce widely different part characteristics in terms of strength, aesthetics, cost and so on. Close vendor/supplier cooperation is not always possible of course, especially at the conceptual stage, but some knowledge of possible strategies, potential materials, as well as a clear part function specification will be very useful.

There are some general guidelines that should be considered when designing parts that might potentially be made using AM processes. One of the most important things to remember about most AM parts is that they may be heavily **anisotropic**, their properties may differ considerably depending on the direction in which analysis (of, say, stress) is being considered. This is principally because of layering. At a practical CAD level it means that design orientations are important and re-orientations may be necessary to prepare parts for manufacture. It is worth paying some attention at the outset to the ISO standard methods of referring to axis orientations. Figure 15.2 shows the **ISO 841: 2001** standard axes set up for NC machining in the Z-upwards orientation and this is replicated for AM purposes by ISO. Note that the three rotational axes all follow the '**right-hand rule**', as described in Chapter 11.

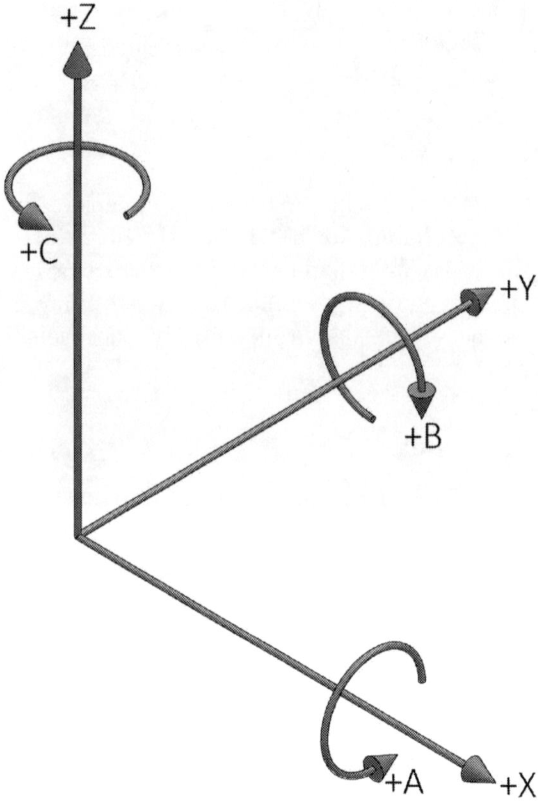

FIGURE 15.2 Z-upwards orientation of axes.

Note that these axes may be shown in other orientations, for example, in a 'Z-downward' manner; however, the relationship between all six coordinate directions remain fixed. The general rule for AM is that the Z direction should point in the direction of material addition but it is worth checking with any manufacturer what the axis regime is for a particular machine in case it doesn't follow the ISO system. In some applications the axes may not lie in horizontal/vertical alignments, with material being added at angles other than 90° to the xy, xz or yz planes. This is sometimes the case with jetting processes. Always check.

Some general design rules follow.

15.4.1 Holes

Due to the anisotropic nature of AM processes it is generally good practice to try and orientate holes parallel to the Z axis as this will maximize their hoop strength. Small holes (1 or 2 mm or less) may also cause problems simply due to uncertainties about process precision. In such cases it may be necessary to add a secondary post-process operation to drill these out for a more accurate size. Holes requiring threads may present additional concerns for the designer.

15.4.2 Threads

In many 3D CAD systems, thread details are stored attributes on shafts and holes and not actually modelled as solids as this can be computationally expensive. However, AM systems require full 3D depictions of the threads they are to make. Increasingly CAD modellers offer the ability to transform attribute thread descriptions into full 3D modelled forms. This can be useful, especially for larger thread sizes. For smaller or more precise work it may be necessary to consider using postprocesses to cut threads or to use **studs** or **inserts** for threads.

15.4.3 Edges

Edges and corners on machined components often need deburred or chamfered or even rounded, although rounding is often avoided for cost reasons. AM processes, on the other hand, do not incur large costs to round edges and this should be considered, especially when it might mean further finishing operations might be avoided. Sharp edges are generally to be avoided with many AM processes, except when the edges have specific functional roles.

15.4.4 Wall Thickness

Thin walls can cause problems with breakthrough or simple lack of strength, especially where good alignment and bonding between layers is likely to be difficult. In general, depending on the process, however, when wall thickness to wall height ratios are less than 1:20 or so it may be worth adding support structures such as **webbing** or **cross members** between walls.

15.4.5 Internal Structures

It is common—often to save material and achieve lighter part designs—that designers might opt to fill large volumes with internal structures such as **cross webbing** or other **cellular patterns**. Care has to be taken to ensure the required strength of course, and analysis can be complex. A second and important consideration is to ensure that unused raw material such as liquid or powder is not trapped in the finished part. This requires analysis of the part's in-process geometry as well as its finished state and this can also be a matter of concern for part stability during manufacture.

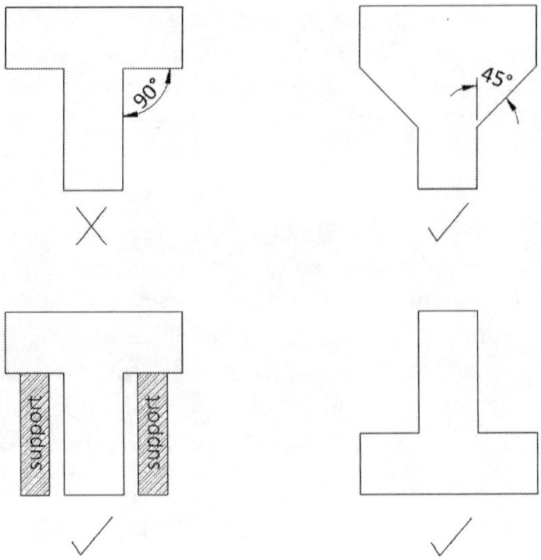

FIGURE 15.3 In-process part stability.

15.4.6 In-Process Shapes/Orientation and Part Stability

During AM processes it is possible that a part will become unbalanced or that the weight of large overhangs will compromise the dimensional integrity of a part. In general, overhangs that are around 45° or less to the vertical are often acceptable. For values greater than this, for example, at 90° to the vertical, it may be worth considering redesigning, adding supports or changing to a more favourable part orientation as shown in Figure 15.3.

Some processes do offer a certain amount of support naturally, for example, powder-based methods.

15.4.7 Unused Raw Material Removal and Disposal

As mentioned earlier, removal of any unused raw material can be a problem when using AM and this material may have limited scope for recycling. Unused material may be partially processed, for example, melted or contaminated, and many of the raw materials used in AM will require specialized and expensive disposal, especially where these are poisonous.

15.5 FILE TYPES

The most common file type for use with AM work is STL, and second to this is OBJ (see Chapter 11 for descriptions). Two further file types that are used and accepted by AM manufacturing companies are AMF and 3MF. Table 15.1 shows a brief comparison of the file types and their applicability—in the opinion of the author.

TABLE 15.1
AM File Type Comparison

	STL	OBJ	AFM	3MF
Popularity	Very high	High	Low	Medium
Level of detail	Poor: flat triangular meshes approximate surfaces and are prone to error	Excellent: wide range of geometry options including non-triangular mesh and splines	Excellent: curved triangular meshes	Excellent: curved triangular meshes
Colour/texture	No	Yes: in a separate file	Yes	Yes
Material	No	Yes	Yes	Yes
Meta-data	No	No	Yes	Yes
Comments	• Good for single colour and prototyping • Compatible with nearly all CAD and manufacture • CAD designed parts generally produce better STL files than those from scanning devices (which may require complex error correction processes)	• Some of the geometric options may not be implemented within CAD systems, for example, the ability to store accurate spline data	• Compatibility may be limited due to slow takeup	• Compatible with a reasonably wide range of systems • Owned by a small group of companies

16 Surface Texture

16.1 INTRODUCTION

Dimensions shown in typical engineering drawings normally relate to relatively 'large distances' of several millimetres or more. The tolerances that are shown in drawings relate to boundaries that contain an edge or surface; however, between these limits the features are undefined. Comments can be added to drawings (e.g. 'NOT CONCAVE'), but these can be ambiguous. Sometimes we might want to characterize the acceptable shape of a surface's fine detail as it might be necessary to meet a particular design intent. We may, for example, want to specify how rough or how spikey a surface should be or what the spacing might be between peaks in periodic waveforms where these are present.

The possibilities for describing the shapes of surfaces are endless and beyond the scope of this book; however, the ISO standards describe a limited but useful variety of characterization methods that might be used to limit **surface texture** variability for a designer. The standards **ISO 21920–1: 2021, ISO 21920–2: 2021** and **ISO 21920–3: 2021** describe **profile-based** methods, while **ISO 25178–1:2016, ISO 25178–2:2022** and **ISO 25178–3:2012** give an outline of **areal**, or area-based, methods. Further areal detail, such as the characteristics of different measuring machines and software standards, are given in the various other individual **ISO 25178** standards that relate to specific measuring technologies or to measurement software standards and file formats. At the time of writing there are a further 14 such standards.

Note that these standards replace the now-withdrawn **ISO 4287** standard.

Commonly, relatively simple profile-based **amplitude** parameters are used to characterize surfaces in everyday manufacturing situations. However, there are many specialist situations and applications that require more sophisticated measures, for example, where the **height distribution** of material is very important for specifying bearing surfaces (it has a considerable effect on both friction and lubrication retention) or where aesthetics and/or tactile characteristics are important (e.g. computer touchpads), and in such cases very rigorous and complex areal-based specifications for surface texture tolerances might be expected.

What follows is an overview, as defined by the ISO, of surface characterization methods and their measurement as well as descriptions on how these should be shown on engineering product communications including drawings and PMI.

16.2 PROFILE VERSUS AREAL METHODS

Traditionally, surface roughness was specified using some form of profile method. For example, an **average roughness** value might be given on a drawing, and this would be based on the average amplitude of deviation from an ideal or mean

DOI: 10.1201/9781003395072-16

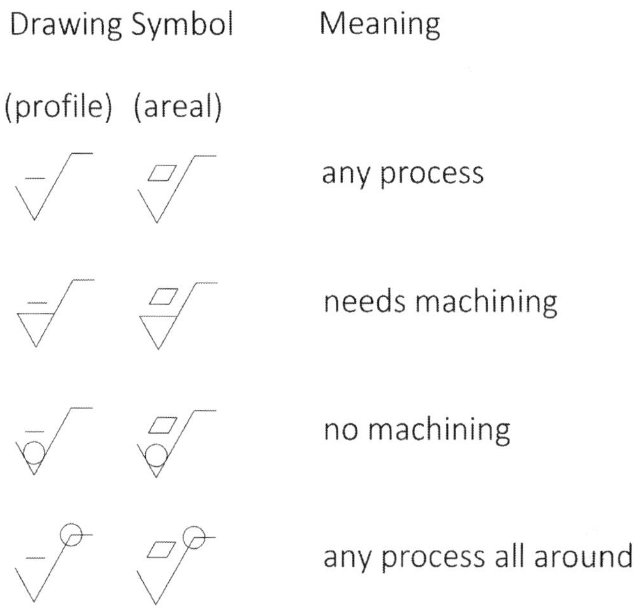

FIGURE 16.1 Basic symbols.

profile. In many cases, this system works very well, but advances in the science of roughness, its specification and measurement, have also made areal methods of characterizing surface texture increasingly popular. These methods are based on scans of areas on parts, and while it is possible to compute these using a large number of profiles, they are also commonly measured with optical instruments that can assess an entire area.

In general, there are three indicators that have traditionally been used on drawings to show that a particular surface texture characteristic is required. These, along with the 'all around' option, are shown in Figure 16.1. Examples of their use will be shown once some of the definitions of common parameters and symbols that might be included in these are outlined.

There are a number of ways these symbols can be used, and with a number of options for specifying parameters, these differ between profile and areal methods. The symbols may refer to a profile (edge) or surface (in 3D) in an engineering drawing by being attached to extension lines, leader lines or to tolerance indicators (feature control frames).

16.3 PROFILE METHODS

On an engineering drawing, many surface texture characterizations rely on the use of roughness parameters that are based around profile measurements, for example, height measurements that are perpendicular to the line along which a sample is taken or spacing measurements that relate to the distance between irregular profile features. Profile methods can also be used to specify and measure hybrid characteristics such as local profile gradients or to show material distribution graphs. There are a number of common parameters that can be used to specify desired surface characteristics of these types, and they can be validated in a number of ways. One simple method is with the use of **comparator sets**. These consist of a collection of surface samples that have been prepared with known roughness values and classified into broad categories. A user can feel these with their fingers and then compare their tactile observation to that on the target workpiece and hence judge into which **roughness classification** (typically an **N1–N12** scale) a surface should be declared to belong.

More commonly, **profilometers** are used and these are usually based on a **stylus probe** mounted on a pick-up arm (similar to that on a hi-fi). The arm is dragged across the workpiece as indicated in Figure 16.2.

Naturally, the mechanical aspects of these measures impose some **filtering** of the surface shape profile. This is due to characteristics such as the stylus radius and the dynamics (**frequency response**) of the arm and stylus sensor arrangement. Sometimes the stylus arm is fitted with a skid, which filters out some

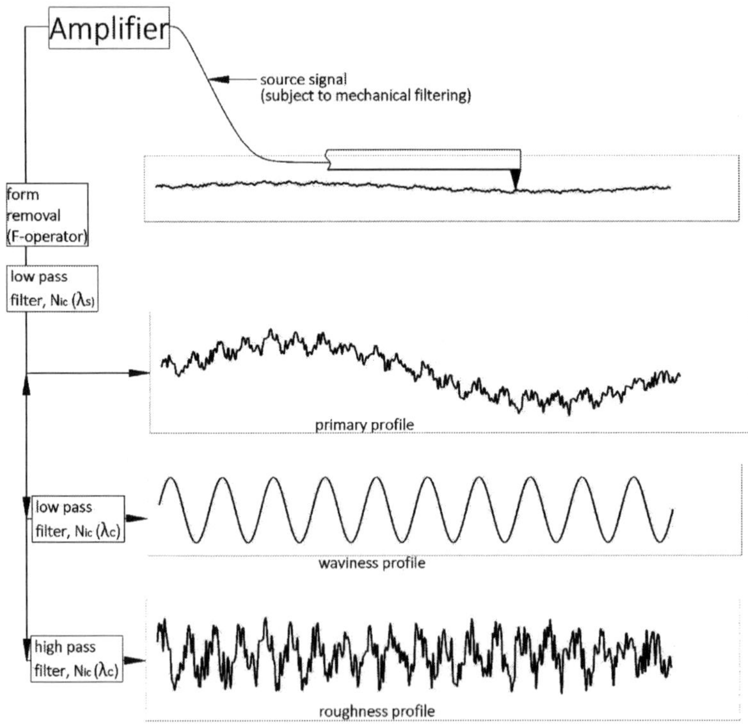

FIGURE 16.2 Typical stylus/profile-based surface texture measurement arrangement.

unwanted low frequencies. The length of the sample taken must also be controlled to ensure that any desired frequencies or estimates of height are sufficient to give reliable test results.

16.4 FILTERING

Although the signal which comes out of an electro-mechanical pick-up device has already been subject to some mechanical filtering, there are further filters that can be applied using graphical techniques and/or analogue and/or digital electronic devices. Traditionally, when performing graphical analysis, filtering could be applied by limiting the lengths of samples that were being used to evaluate parameters. Shorter lengths would filter out unwanted low frequency data. Analogue electronic filters or digitally sampled data can similarly limit the signal frequencies to those of interest to the designer and metrologist but these should be carefully controlled and hence specified on drawings/PMI. The ISO standards give detailed descriptions on how this should be done.

Generally the process of filtering profile data might begin with some form of **low-pass filter** either imposed mechanically or explicitly stated through the use of what is termed and **S-filter** (or short wavelength filter) of value lambda-s, usually stated in µm units. An **F-operator** is also commonly applied which removes **form elements** from a profile signal. This can involve fitting predetermined shapes (splines or geometric primitives such as least square mean lines) to the data so that these basic shapes or forms can be removed from the profile signal. Additionally a **high-pass filter** might be applied to filter out long wavelengths from the profile and this is done by specifying the values for an **L-filter**, denoted by the **nesting index N_{ic}**. The term **'cut-off'**, is often used to specify a wavelength (λ_c) at which 50% of a signal will be attenuated. The **ISO 21920–3**:2021 standard gives preferred cut-off values for λ_c of 0,08 0,25 0,8 2,5 and 8 mm for profile parameters and ISO 25178–3:2012 gives preferred cut-off values (N_{ic}) of 0,1 0,2 0,25 0,5 0,8 and 1 mm for areal parameters. Strictly speaking, the term 'cut-off' really only applies to linear filters, but there are other non-linear ones defined in ISO standards (e.g. ISO 16610) and the term 'nesting index' (N_{ic}) is more meaningful for these.

16.4.1 Amplitude Parameters

The most commonly used methods of characterizing the roughness of a workpiece use **'R-based'** (roughness) measures. These can use field parameters (calculated over a full evaluation length) such as R_a (also formerly known as **CLA**), R_q (or R_{rms}), R_{sk}, R_{ku} and R_t, or they can use feature parameters (calculated over a given number of section lengths and then combined) such as R_z. These measurements are based on the output from filters which may be specified explicitly or using default settings as described in the **ISO 21920–3:2021** standard and which are normally available as options on most measuring devices. The various measurement conditions under which a workpiece can be evaluated are grouped into specific setting classes (**Sc1, Sc2, Sc3, Sc4** or **Sc**5), and each will specify a regime of high-pass filter wavelength, low-pass filter wavelength, the minimum distance between samples and the maximum permitted stylus tip radius, the full evaluation length and the maximum stylus tip radius.

Note that values for low pass, or primary profiles, for example, P_a, P_q, P_{sk}, P_{ku} and P_t as well as profiles that have been bandpass filtered to isolate waviness, for example, W_a, W_q, W_{sk}, W_{ku} and W_t, can all be found using the same formulae for the following high-pass *R*-based values.

16.4.1.1 R_a—Roughness Average, R_q—Roughness Root Mean Square, R_{sk}—Roughness Skew and R_{ku}—Roughness Kurtosis

The R_a value is simply the average deviation from the surface mean line. For a given profile over a specific evaluation length, the mean line can be determined as the overall average value of the profile height. In analogue electronic measuring devices, this can be done using a very low-pass filter over a limited sampling length. The R_a is then simply the average deviation from the mean line to the signal samples. This is shown graphically in Figure 16.3.

$$R_a = \frac{1}{n}\sum_{i=1}^{n}|z_i|$$

or in analogue terms (as defined in **ISO 21920–1:2022**):

$$R_a = \frac{1}{L}\int_0^L |z|\,dx$$

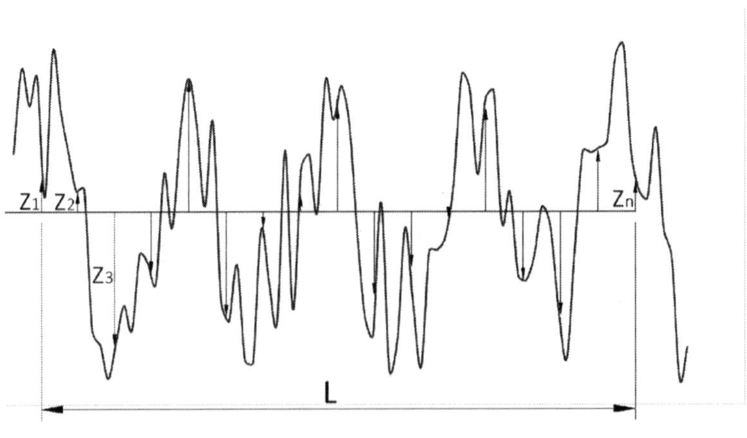

FIGURE 16.3 Measurement of deviations from mean line.

where z is the profile height above (+ve) or below (−ve) the mean line and L is the evaluation length.

If we wished to get a primary profile (P-based) or waviness profile (W-based) version of the previous equation, we can simply substitute these into the R-based formula as follows (but the z values will be based on different filtered data):

$$P_a = \frac{1}{L}\int_0^L |z|\,dx$$

or

$$W_a = \frac{1}{L}\int_0^L |z|\,dx$$

The use of R_a to specify a roughness tolerance is very common, particularly in certain regions, for example, the UK. Comparator surface sets are readily available for validation of manufactured parts. These, and nearly all electronic surface texture measuring devices, will offer the measurement of R_a.

It is quite common practice for designers to specify a particular R_a value on its own, and in doing so they are accepting the use of a set of default conditions for the measurement of the workpiece that will validate the tolerance. For example, if an R_a value is specified with no further information **ISO 21920–3:2021** assumes the following general defaults:

- The measurement instrument will be a mechanical device (or simulate one).
- The direction of measurement will be the one that yields the maximum value, usually normal to the surface lay (see descriptions later in this chapter).
- No measurement should exceed the tolerance value (see tolerance acceptance rules later in this chapter).
- The stated tolerance is an upper tolerance limit (i.e. with no minimum tolerance value).

In addition, there are specific defaults for certain parameters; for example, if 'R_a 1' is specified, this falls in the $0{,}1 < R_a <= 2$ category, and therefore specific default values are:

- Cut-off length = 0,8 mm (the high-pass wavelength filter setting λ_c).
- Five sample lengths (or 'sections') of 0,8 mm should make up the evaluation length of 4 mm.
- The measuring stylus should have a maximum tip radius of 2 µm.
- The maximum sampling distance between measurements should be no more than 0,5 µm.

ISO 21920–3: 2021 details all of the default values for the various profile parameters, and these are normally acceptable for everyday use; however, if a designer needs to specify other values of measurement settings, then this can be done in the surface texture tolerance indicator on the drawing/model.

Where two separate surface texture parameters are specified in the surface tolerance indicator, then the upper or first shown parameter values will take precedence.

Although it is most common to specify the upper tolerance limit, for example, the maximum degree of average roughness tolerable, it is also useful to be able to specify bi-lateral tolerances with both minimum and maximum permitted tolerance values. These use different default values which are based on the central value halfway between the two tolerance limits.

Using the *z* values of the sample or signal, we can also calculate several other parameters as follows:

$$R_q = \sqrt{\frac{1}{L}\int_0^L z^2\,dx}$$

$$R_{sk} = \frac{1}{R_q^3}\frac{1}{L}\int_0^L z^3\,dx$$

$$R_{ku} = \frac{1}{R_q^4}\frac{1}{L}\int_0^L z^4\,dx$$

The R_q or 'root mean square' measure is also commonly used as a measure of the dispersion of a surface profile about the mean. The R_{sk} value measures the skewness of the surface profile and the value can either be positive or negative. A negative skew is generally better for bearing surfaces, and engineers often specify R_{sk} values of around –1,6 to –2 for such surfaces. Figure 16.4 shows two surfaces, one negatively skewed and one positively skewed.

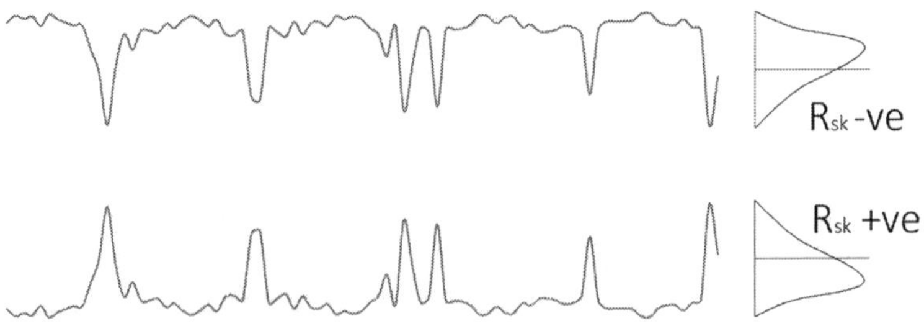

FIGURE 16.4 Skewed profiles.

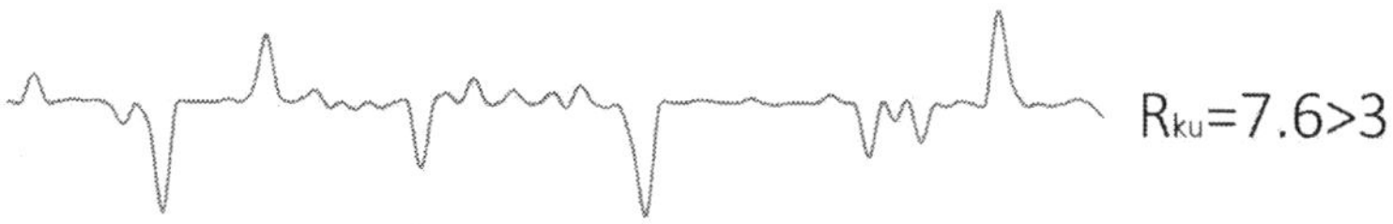

FIGURE 16.5 Profile with high kurtosis.

The R_{ku} value is a measure of kurtosis, or 'tailedness', of the distribution. This is a measure of the importance of outliers, that is, extreme values on the profile. If the profile follows a normal distribution we would expect an R_{ku} value of 3; however, manufactured surfaces may not follow a normal distribution; for example, we might expect sinusoidal shapes (R_{ku} = 1,5) or other shaped profiles like square, triangular or random forms. Figure 16.5 shows a curve that exhibits a high level of kurtosis.

It is worth remembering that the R_{ku} value here is based on the definition given in **ISO 21920–2:2021**. Some other measures are in use which seek to calculate a kurtosis value that would lead to a normal distribution yielding a value of 0 (roughly R_{ku}–3 for large datasets). For example, this is the default measure used for kurtosis functions in spreadsheets such as Excel.

16.4.1.2 R_t—Total Height

The total height can be defined as the difference in height between the largest height value and the lowest depth value in an evaluation profile.

$$R_t = z_{\max} - z_{\min}$$

Figure 16.6 gives a graphical representation of the previous formula.

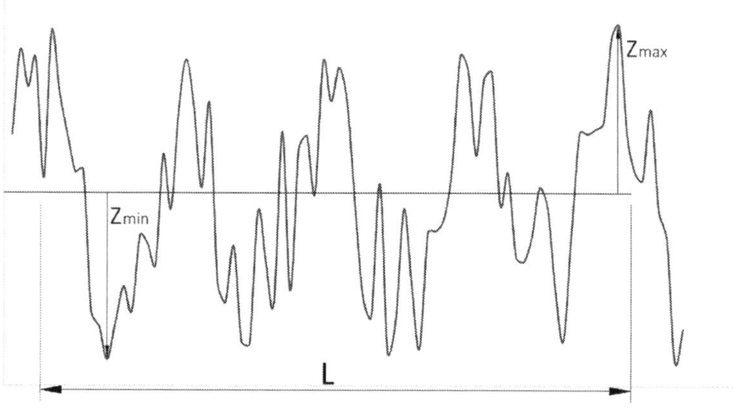

FIGURE 16.6 Total height measure.

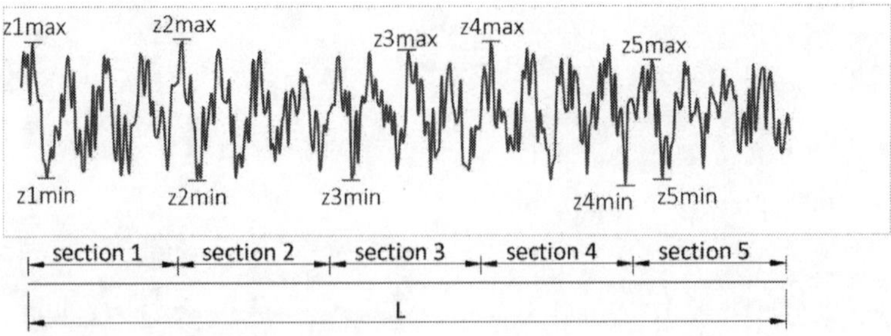

FIGURE 16.7 Maximum height measure.

16.4.1.3 R_z—Maximum Height

The maximum height parameter is what is referred to in the ISO standards as a feature-based parameter or section length parameter. These are not calculated along the whole evaluation length but consist of a combination of measures taken in each section (or sample). For example, R_z is the average of the highest peak to lowest valley measurement for each of the (normally 5) sections that make up the evaluation length as follows:

$$R_z = \frac{1}{n}\sum_{i=1}^{n}(z_i \max - z_i \min)$$

This is shown graphically in Figure 16.7.

16.4.2 Spatial Parameters

Two spatial parameters are defined in the ISO standards, the autocorrelation length, R_{al} and the dominant spatial wavelength, R_{sw}. These parameters are difficult to compute and are not commonly employed in everyday use. Where a design requirement arises for a specific value of either of these measures, it is likely to be the result of a complex analysis of the surface function whereby the validation process will be known beforehand. Calculation of the actual values of these parameters may be difficult to carry out, as they involve calculating autocorrelations and Fourier modelling, from a sample profile data set; however, they may be available as functions in some profile measurement devices.

16.4.3 Hybrid Parameters

Hybrid parameters are based on local gradient values of profiles. One popular measure is the root mean square gradient parameter R_{dq}, which is defined as follows:

$$R_{dq} = \sqrt{\frac{1}{L}\int_0^L \frac{dz_i}{dx}^2 dx}$$

or in discrete terms,

$$R_{dq} = \sqrt{\frac{1}{n}\sum_{i=1}^{n}\left(\frac{\Delta z_i}{\Delta x}\right)^2}$$

Other parameters such as the arithmetic mean of absolute gradient, R_{dt}, and the developed length, R_{dl}, may also be calculated in a similar way as the profile parameters but based on the derivatives as in the previous example.

Similarly, as before, we can also define values for the unfiltered data, for example, P_{dq}, P_{da}, P_{dt} and P_{dl} or for the waviness data, W_{dq}, W_{da}, W_{dt} and W_{dl}.

16.4.4 Tolerance Acceptance Rules T_{MAX}, T16% and T_{MED}

The **ISO 21920–1:2021** defines three alternative rules for validating a workpiece surface that is subject to a surface texture tolerance. By default, no measured value of roughness should be allowed to lie outside of a specified tolerance but this can be explicitly stated, if necessary, by the inclusion of the following symbol within the roughness specification:

T_{max}

meaning no measured value can exceed the tolerance limit.

On occasion, it may be desirable to accept that some measured parameter values be allowed to lie outside a stated tolerance, and ISO defines a specific value of 16% as a standard option that can be included on a roughness specification. The symbol for this acceptance rule is:

T16%

meaning that 16% of the number of measured values can exceed the tolerance limit.

A further acceptance rule that allows for the median value of a group of measures (e.g. where multiple samples are used) to lie within a stated tolerance limit is specified using the following symbol:

T_{med}

The use of this rule is not common in general machining environments.

16.5 AREAL METHODS

When using an areal specification, the basic surface measurement symbol should be added to the texture requirement indicator, as shown in Figure 16.1.

Areal methods of characterizing and validating surface texture are largely based on those that were developed for profile-based parameters. The main difference is of course that areal methods are based on defined areas of surfaces and as such are commonly used in the context of surface measurement instruments which scan entire areas such as the many modern optical devices that are available.

As with profile methods, areal field parameters are available and are directly analogous to their profile counterparts although the surface measurements often exhibit larger deviation values than the profile measurements. The height-based parameters S_a, S_q, S_{sk} and S_z, for example, are available.

16.6 LAY

Surface lay descriptions relate to the predominant direction of the surface texture patterns; they depict in which direction the prominent grooves run. Usually we wish to measure surface roughness normal to this direction because it will most likely be the orientation that leads to the greatest roughness value but other directions can be used. The direction of the surface lay that should be evident on the workpiece, and the direction of roughness measurement in relation to the surface lay can both be shown in the texture requirement indicator. **ISO 21920–1:2021** gives directions on how to use descriptions of lay direction on drawings/PMI for profile-based use, and **ISO 25178–1:2016** describes the areal guidelines. These are essentially the same, but at the time of writing **21920** is the more up to date and gives examples using orientation indicators.

Typical lay patterns might be as shown in Figure 16.8.

Lay Symbol	Description	Example
M	multidirectional	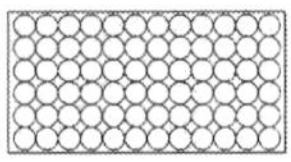
C	circular	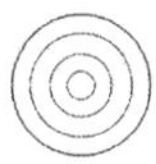
R	radial	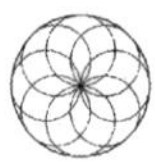
P	particulate	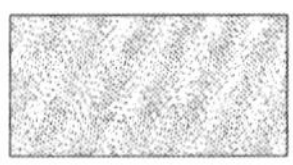
=	parallel	
⊥	perpendicular	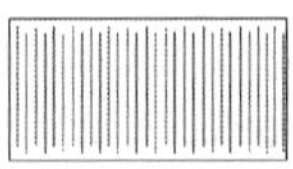
X	cross-pattern	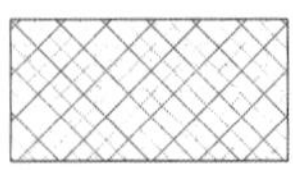

FIGURE 16.8 Basic lay symbols.

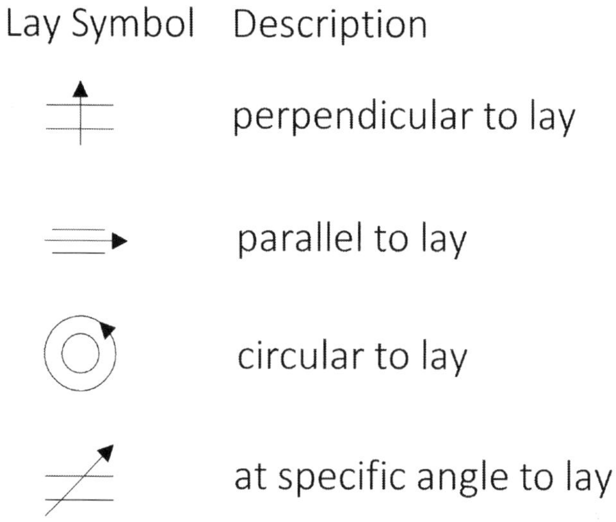

FIGURE 16.9 New symbols showing measurement relative to lay as in ISO 21920–1:2020.

When surface profile measurements are being taken relative to lay directions, these should be indicated using new symbols that are defined in **ISO 21920–1:2020**, as shown in Figure 16.9.

16.7 EXAMPLES

Now that we have described some of the types of texture measurement, common parameters and lay symbols, we now consider some examples of how these can be used on drawings/models.

16.7.1 Minimum Indications

ISO 21920–1:2021 specifies that the minimum information that must be shown for a profile is as shown in Figure 16.10.

Figure 16.10(a) shows the minimum required information required to specify surface texture on a drawing/PMI where parameters have defined defaults as shown in **ISO 21920–3: 2021**, where a = parameter specification and b = parameter value. Figure 16.10(b) shows the minimum required information required to specify surface texture on a drawing/model for parameters with no default values where c = specification of measurement values. Specific examples follow.

16.7.2 Example 1—Minimum Requirements

Figure 16.11 shows minimum requirement indicators with some example specifications.

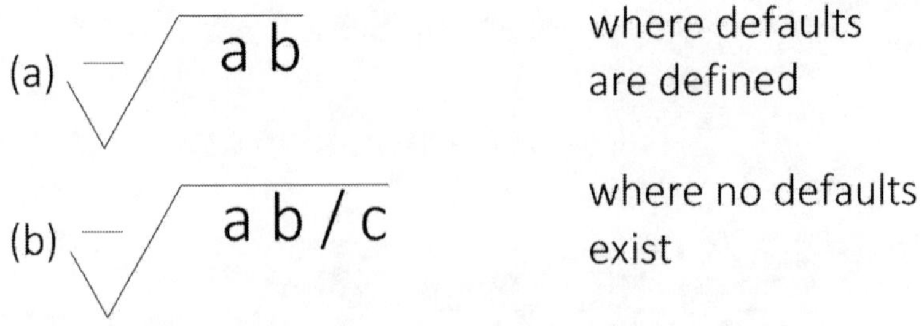

FIGURE 16.10 Minimum required information for texture specification.

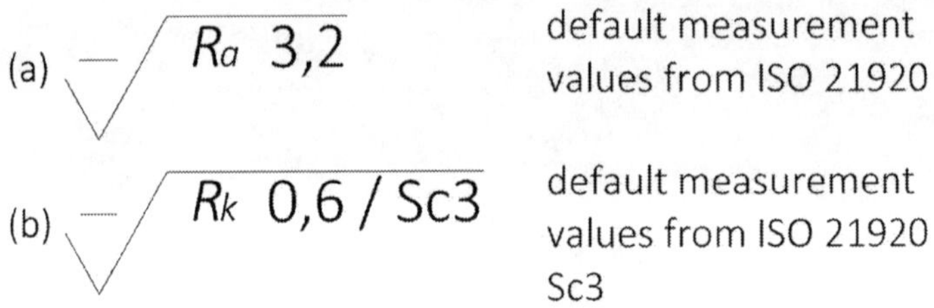

FIGURE 16.11 Example of minimum required information for texture specification.

In Figure 16.11(a) an R_a value of 3,2 is shown. As no other information is given, the general default values as outlined in 16.4.1.1 should be assumed and that the parameter shown is an upper tolerance value. For the specific value of $R_a = 3{,}2$ this would additionally mean the defaults will be defined by class setting Sc4, and correspondingly that the L-filter or cut-off value (λ_c) should be 2,5 mm, the evaluation length should be 12,5 mm, the maximum sampling distance should be 1,5μm and the stylus tip radius should be no greater than 5 μm.

In Figure 16.11(b) an R_k value of 0,6 is shown. R_k is a parameter that is used primarily in the French motor industry for specifying cylinder bore linings and has no default measurement details defined in **ISO 21920** so a specification class of Sc3 is given. This means again that general defaults can be used and specific values for Sc3, and correspondingly that the L-filter or cut-off value (λ_c) should be 0,8 mm, the evaluation length should be 4 mm, the maximum sampling distance should be 0,5μm and the stylus tip radius should be no greater than 2μm.

16.7.3 Maximum Specification Options

The maximum profile-based specifications that can be applied to a surface texture indicator, as defined in **ISO 21920–1: 2021**, depend on whether the profile parameter being specified is a field or feature parameter and on whether it is *R*-based, *W*-based or *P*-based. An example of the information which might be entered for an *R*-based field parameter specification can take the form shown in Figure 16.12.

Note that the previous symbol can be augmented with orientation operators for both lay direction and profile measurement direction, and these may be helpful in 3D PMI specifications.

The basic texture requirement indicator for areal-based indicators, as outlined in ISO 25178–1:2016, is used as shown in Figure 16.13.

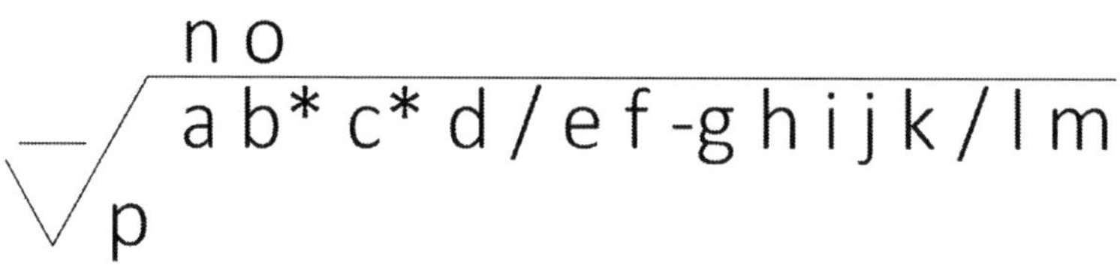

where
a = tolerance type (U, L...)
b= R parameter symbol (Ra, Rz.....)
c = tolerance acceptance rule (Tmax, T16% ...)
e = S-filter type (eg -G = Gaussian)
f = S-filter nesting index, N_{ic} (0,08 0,25 0,8....)
g = L-filter type
h = L-filter nesting index
i = evaluation length (0,4 1,25 4...)
j = F-operator association method and element (-G, S,...)
k = F-operator nesting index, N_{ic}
l = profile extraction method
m = placeholder for additional information
n = manufacturing process
o = type and direction of lay (M, C....)
p = measuring profile orientation in relation to lay
* = compulsory field

FIGURE 16.12 Full specification options for evaluation-length *R*-based parameters.

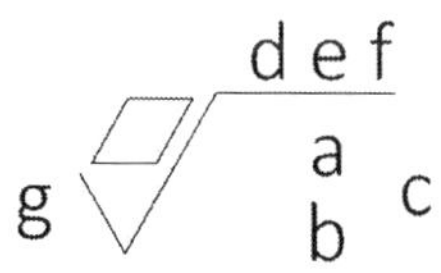

where
a is the primary roughness parameter specification
b is the secondary roughness specification
c is the intersection plane showing orientation of measure
d is the manufacturing method / coatings etc
e is the direction surface lay pattern
f is the intersection plane showing orientation of lay

FIGURE 16.13 Areal symbol positions.

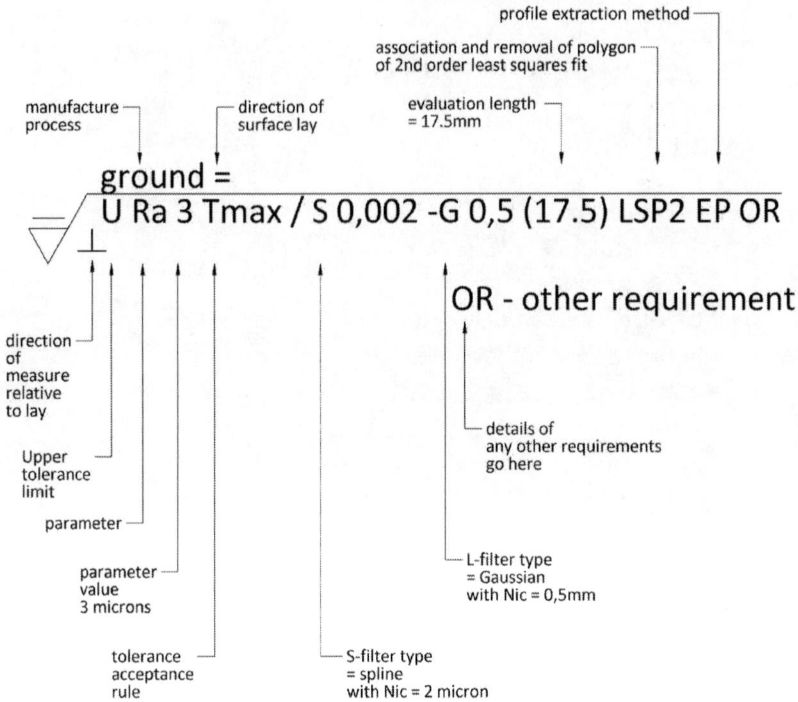

FIGURE 16.14 Example of full specification for evaluation-length R-parameter.

16.7.4 Example 2—Full Specification

In the example shown in Figure 16.14 the maximum specification is used to show explicit defaults and non-default values of the measurement regime.

16.7.5 Example 3—Positions of Texture Indicators on Drawings

The example shown in Figure 16.15 shows the texture indicator being placed in a variety of positions, for example, on edges, on extension lines and on feature control frames. Note also the use of multiple indicators to show specific machining requirements.

16.7.6 Example 3—Use of 'All-Around' Indicator

The example shown in Figure 16.16 shows the use of the all-around symbol on the indicator. It is shown attached to a leader line and the specification applies to the surfaces 1–6. Note that for these surfaces no machining is allowed. The remaining two surfaces are not included in the all-around specification and have their own individual tolerances and machining is permitted but not mandatory.

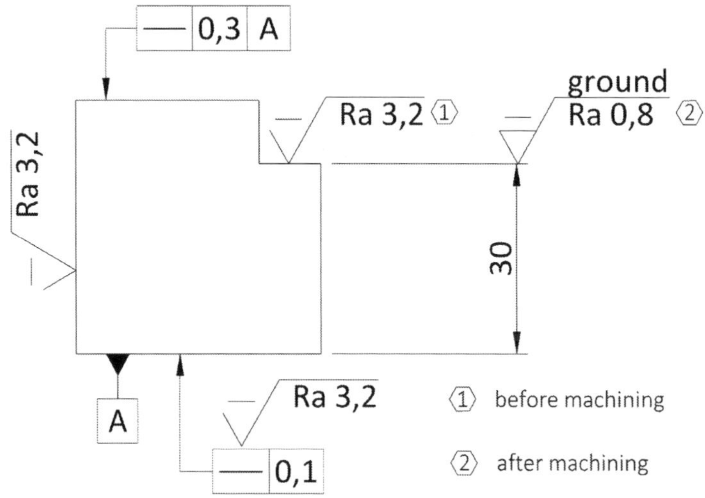

FIGURE 16.15 Example positions of texture indicators on drawing.

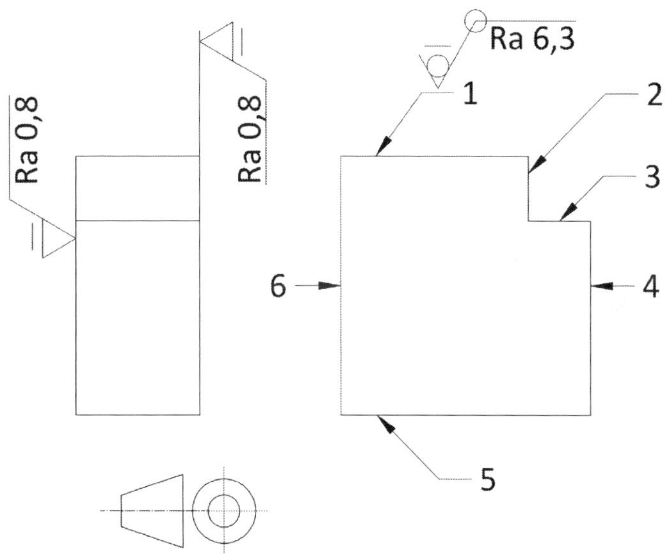

FIGURE 16.16 All-around symbol being used for non-machined surfaces.

16.8 COMMENTS

The chapter has shown some of the complexity that can be involved in specifying surface conditions; however, in most machine shops only a few of the many options will be in everyday use. The last example shows usage of, for example, the R_a symbol.

17 Digital Product Definition, Drawings, 3D PMI and the Digital Twin

17.1 INTRODUCTION

There are many ISO standards that relate to how the output from CAD models should be formatted for presentation to an engineering project community. These give advice on nearly every aspect of presentation, from drawing sheet sizes, line types, title boxes, dimensioning and tolerancing and the preparation of interactive 3D models.

17.2 SHEET SIZES

While 3D views and interactive modelling are revolutionizing design and manufacture practices, it's also true that 2D drawings are unlikely to disappear any time soon. **ISO 5457** (under review at the time of writing) gives instruction on recommended sheet sizes for drawings and on the preferred line types to be used on these. The drawing sheet sizes are based on **A4** (portrait and landscape), **A3, A2, A1** and **A0** sizes, for example, 297 × 210 mm for standard A4 landscape trimmed sheets. The use of **borders** is recommended on all sheet sizes and these typically are indicated by the addition of a drawing **frame** within which all drawings should be contained. For all sheet sizes the left border should be 20 mm wide and the other borders 10 mm wide.

To help locate particular parts of a drawing it is common in many environments to add a **grid referencing** system on the drawing frame. Table 17.1 shows the ISO recommended number of such fields for the various sheet sizes.

The grid labels should be shown in the border areas with numbers in top and bottom and letters (not I or O) in left and right borders. A4 sheets, however, should only use top and right positions. In every case the letters should be shown with vertical orientation and lettering should be 3,5 mm high.

TABLE 17.1
Number of Fields for Grid Referencing

	A4	A3	A2	A1	A0
Long side	6	8	12	16	24
Short side	4	6	8	12	16

DOI: 10.1201/9781003395072-17

No	Example	Explanation	Typical uses
01.1		continuous, narrow	dimension, extension, leader, hatching lines, termination of interrupted views
01.2		continuous, wide	visible edges and outlines, thread crests, section arrows
02.1		dashed, narrow	hidden edges and outlines
02.2		dashed, wide	permissible areas of surface treatment
04.1		long dashed, dotted, narrow	centre lines, lines of symmetry, p.c.d.
04.2		long dashed, dotted, wide	position of cutting planes, application of tolerance areas
05.1		long dashed, double dotted, narrow	Outline of adjacent parts, extreme positions of moving parts, initial outline prior to forming, projected tolerance zones, outline of datum target areas

FIGURE 17.1 Common line types.

17.3 LINE TYPES

ISO 128–2:2022 describes how **lines** should be depicted on drawing sheets. This standard is some 78 pages long and I have read it, so hopefully you don't have to! Here's the gist of it however. Lines are classified by a composite number such that a two-digit number identifying the line type is used, followed by a point and single digit number that identifies the line thickness. Figure 17.1 shows the most common combinations used on drawings.

For line types as used in specific application areas you can check if your CAD system adheres to the ISO standards and use these or refer to **ISO 128–2:2022.** Note that the width of a line is partly determined by its specification and its context. Narrow, wide or extra-wide lines should be in the ratio 1:2:4 and for most engineering drawings 0,25 mm (narrow) and 0,5 mm (wide) suffice for smaller sheet sizes or 0,35 and 0,7 mm for larger sheets or where extra emphasis might be necessary to aid clarity if drawings are copied.

17.4 TITLE BOXES

Within the drawing area there is normally a title box that gives information about the provenance of the drawing and other non-geometric information. Many organizations choose to use a standard title box; however, this can lead to large boxes with many unused fields in them and these can limit the usable drawing area. This can be traded off with the complexity of making up several standard title box options or allowing bespoke ones to made or edited from the standard ones. It is always good practice to keep any drawing (or 3D depiction) as simple and clear as is reasonable.

ISO recommendations for title boxes are that they should be located in the bottom right of a drawing for A0 to A3 sizes, all of which are assumed landscape. For A4 drawings bottom right should be used for landscape but the whole bottom frame width can be used for portrait orientations.

There is no set of standard fields from which to choose to use in title boxes but typical ones are;

- Drawing title
- Supplementary title
- Drawing number or reference
- Date of issue
- Sheet number (e.g. '1 of 10', '2 of 6', etc.)
- Document status (in-work, released, withdrawn, etc.)
- Organization name and/or badge
- Revision (the revision history can be shown in a separate table on the drawing)
- Author
- Approved by
- Responsible department

Some organizations also insert information within the title box about the standards in use, although this, along with the projection symbol can be shown outside the title box.

17.5 SIMPLIFIED DRAWINGS

Parts with patterns or other symmetries can be depicted with **simplified drawings**. The techniques used to create simplified drawings were originally developed largely to save time in draughting since they cut out repetitive processes. Ironically when using a CAD system it probably takes more time to create simplified views; however, many draughting personnel like to use them because their use can make drawings less cluttered and hence clearer as well as possibly allowing for some space saving. Earlier in the book we showed some examples where time and space could be saved using single dimensions and tolerance indicators for multiple features with, for example, a '4x' statement in front. We also gave some examples of PCDs and springs with simplified displays. Figure 17.2 gives a further example of where reflection symmetry can be used.

17.6 3D PMI

ISO 16792:2021 gives numerous examples of the use of 3D illustrations with **product manufacturing information** added. Most of the time the illustrations show the same techniques that would previously have been used on 2D views and sometimes traditional views might have been clearer in the sense that they implied information that can be lost in 3D. The most obvious example of this is where a

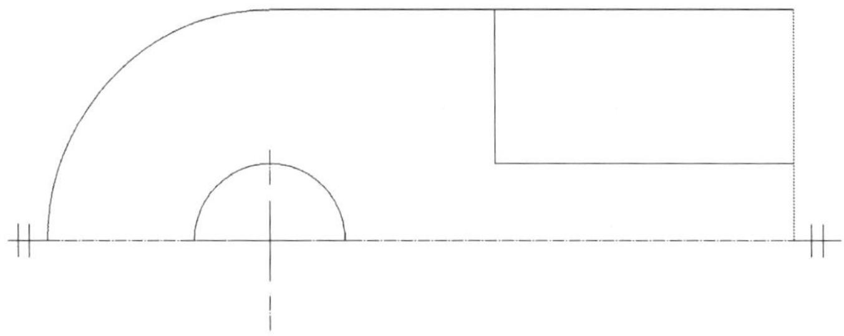

FIGURE 17.2 View simplified with symmetry shown.

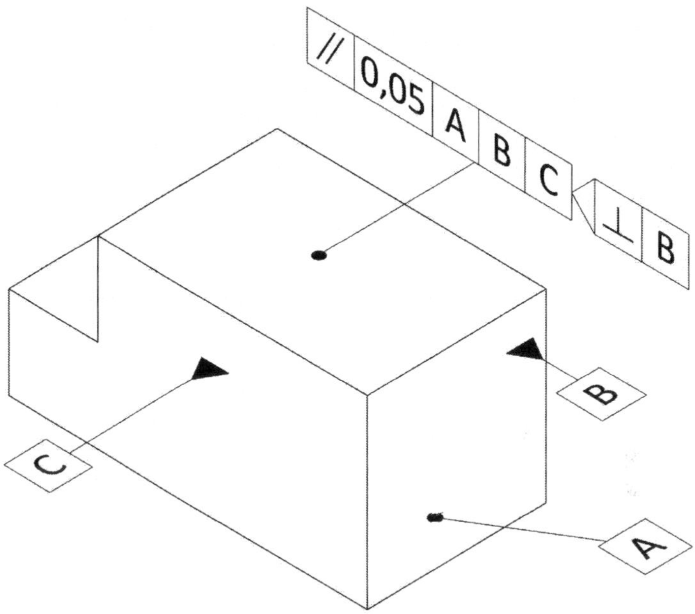

FIGURE 17.3 3D PMI view with direction indicator.

tolerance indicator refers to, say, a straightness requirement and would be shown with the leader line arrow pointing to a line that represents the edge of a plane lying normal to the view direction. This would imply that the tolerance would be validated by taking measurements parallel to the direction of the line.

As shown in Figure 17.3 it would not be obvious in what direction validation measurements should be taken without the addition of an intersection indicator being added to the tolerance indicator.

See also Chapter 7 for other examples of the use of additional indicators to specify validation measurement directions. For reasons of consistency, the ISO standards now mostly recommend that the same techniques should be used in 2D depictions as in 3D. These methods do have the advantage that they offer a richer set of direction indicators for validation and the explicit statements remove considerable room for misinterpretation.

In general, 3D views can appear cluttered and showing dimensions and tolerance indicators on angled planes in 3D can make the values harder to read, so should not necessarily be used in preference to 2D views. Clear 3D images using good practice for adding PMI can help a great deal with visualizing complex topologies, however. They are also particularly handy to add to interactive 3D models where the viewer can control the viewing angle.

17.7 THE DIGITAL TWIN

ISO/IEC 30173:2023 gives an indication of the concepts and terminology that can be used to describe the digital twin across a very wide range of industries. In the realm of manufacture, we might simply think of the digital twin as being any computer representation of a product. This is a very broad definition, and it can include any kind of documentation. Mostly when the term digital twin is used, it implies some kind of lifecycle modelling where the twin would hold information about various stages of a product's life, including any inception documents, contract details and models from the design stage through to decommissioning and disposal. CAD drawings and 3D models are commonly a central part of this and increasingly engineers are using interactive and online models to aid collaboration. In these circumstances the use of 3D PMI models forms an important part of the digital twin.

As explained in Chapter 11 there is no standard that encompasses the entirety of the product lifecycle or the related file formats and structures that might be used to store data, however, many companies are actively developing processes and products that will support activity in this area.

18 Welding

18.1 INTRODUCTION

Welding, brazing and soldering processes often form a major part of the making of engineering components, and as such consideration should be given as to how to specify these in engineering documentation. The material presented here involves a description of the symbols that are used to specify weld requirements and examples of how they might be depicted on engineering drawings/3D models.

18.2 SYSTEM A AND SYSTEM B

As is the case with other areas of engineering documentation, there are two systems in wide operation around the globe which describe welding specifications. These are, as you may already suspect, governed by ISO and by US standards. After many years of trying to amalgamate these systems into one, the present situation is that both systems are acceptable but that only one should be used at a time and the choice should be clearly indicated on documents. The two systems are described by the standards **ISO 2553: 2019** and **AWS A2.4: 2020**, respectively. The AWS standard is published by the American Welding Society. In the following coverage, we will describe what ISO calls **System A**, which is more common with users of the ISO standard, particularly in Europe, and **System B**, which is more widespread around the Pacific Rim countries.

18.3 WELDING PROCESSES

There are many welding processes in use, but every one involves joining materials (usually metals) by placing them together and applying heat so that the joined parts turn liquid or plastic and fuse or diffuse together. Additional filler material is often added either by placing it on the weld area or, more commonly, by feeding it during the welding process from rods or wire feeders. To avoid oxidation of the weld regions during the welding process, the joints may be protected from atmospheric oxygen by adding flux, often coated onto rods, or by flooding with an inert gas such as that used in MIG (metal inert gas) or TIG (tungsten inert gas) welding. A third, more expensive option, is to perform fabrication processes in relative vacuum environments.

18.4 WELDING SYMBOLS

Regardless of the processes that are used, the more common welding types can be depicted on engineering drawings by the use of a system of symbols, as described in **ISO 2553: 2019**. This consists of a line, similar to a break-out line, which is

DOI: 10.1201/9781003395072-18

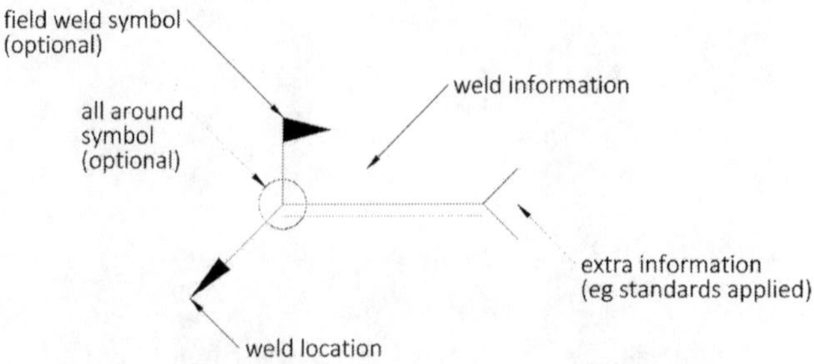

FIGURE 18.1 Basic weld indicator.

augmented with symbols and associated data. What follows is a description of how this line can be applied and an indication of the main symbols and data (e.g. dimensioning and tolerancing) that can be added.

The weld location(s) are pointed to by an arrowed line, which is then broken into a horizontal or vertical reference line to which symbols and data are added. Additional reference information may be added to the tail of the line. This is shown (for System A) in Figure 18.1.

The area labelled 'weld information' contains the dimensions, tolerances and weld type symbol for the particular weld. Further examples later in the chapter will show how the indicator should be used for elementary weld types and for more complex composite welds. What follows is a graphical indication of each of the elementary weld types identified by the ISO standard (on the left) as well as the symbol that should be used on the indicator (on the right). (See Figures 18.2 through 18.25.) The weld regions are shown shaded. Each symbol is also shown with a narrow horizontal line that shows where the symbol should be placed in relation to the indicator line.

18.4.1 Square Butt

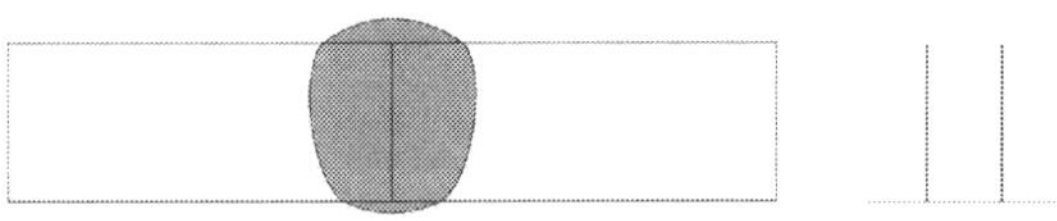

FIGURE 18.2 Square butt weld.

18.4.2 Single-V Butt

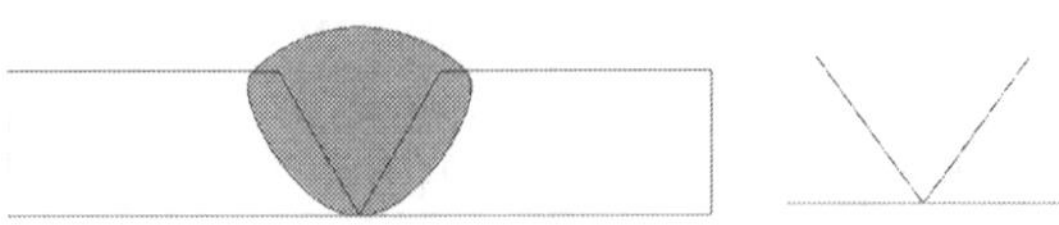

FIGURE 18.3 Single-V butt.

18.4.3 Single-V Butt with Broad Root

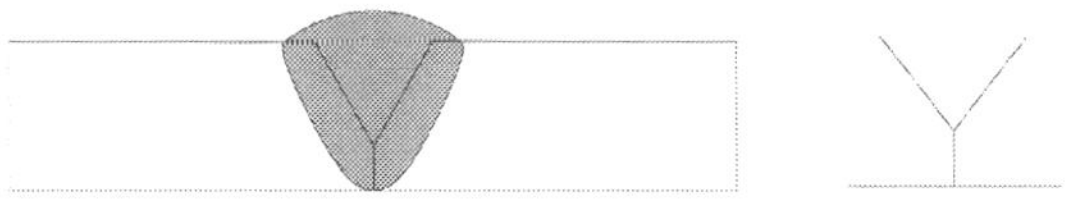

FIGURE 18.4 Single-V butt with broad root face.

18.4.4 Single-Bevel Butt

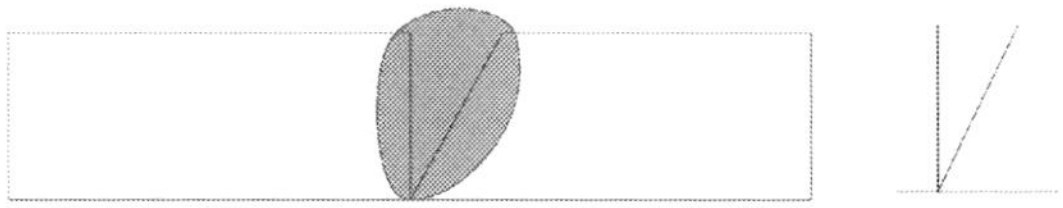

FIGURE 18.5 Single-bevel butt.

18.4.5 Single-Bevel Butt with Broad Root

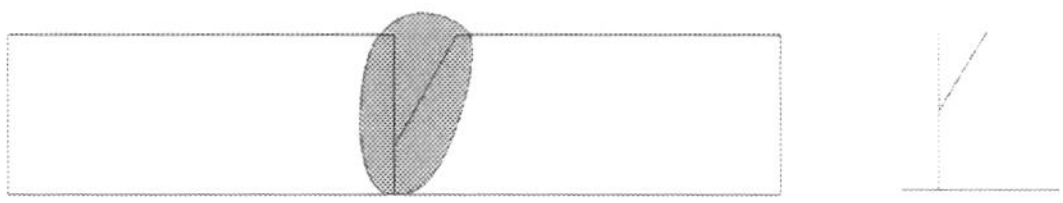

FIGURE 18.6 Single-bevel butt with broad root face.

18.4.6 Single-U Butt

FIGURE 18.7 Single-U butt.

18.4.7 Single-J Butt

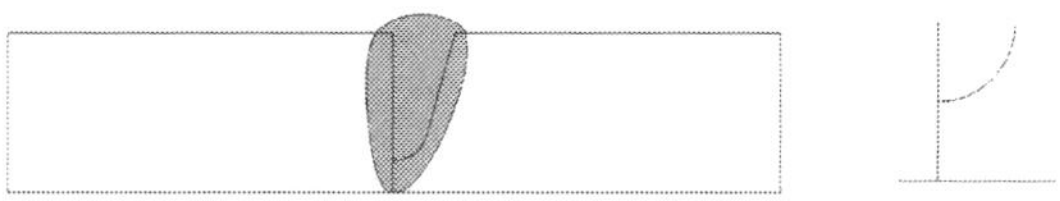

FIGURE 18.8 Single-J butt.

18.4.8 Flare V

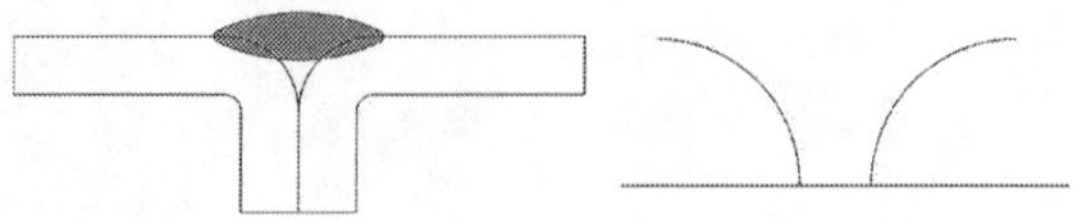

FIGURE 18.9 Flare V.

18.4.9 Flare Bevel

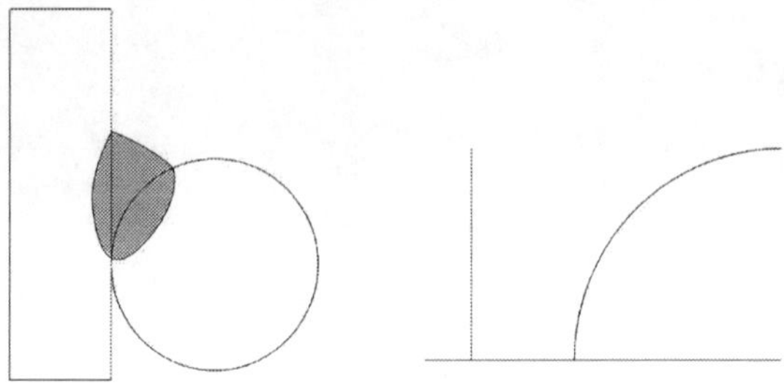

FIGURE 18.10 Flare bevel.

18.4.10 Fillet

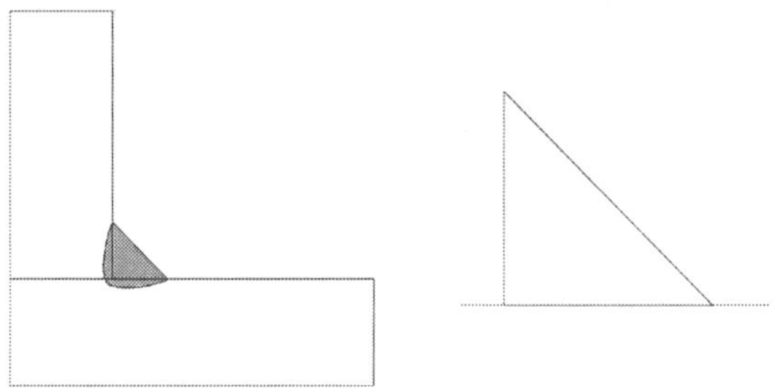

FIGURE 18.11 Fillet.

18.4.11 Plug

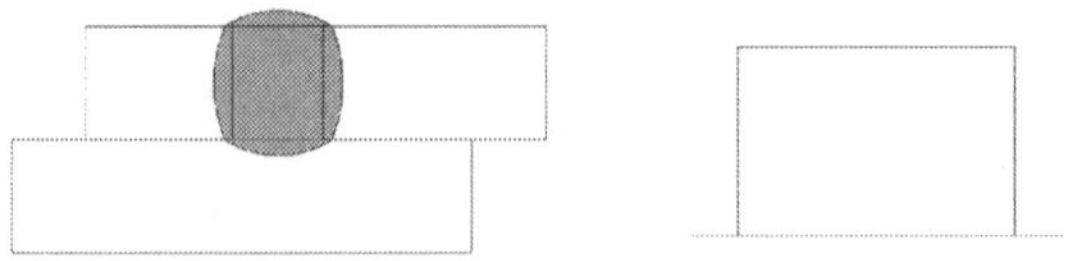

FIGURE 18.12 Plug.

18.4.12 Resistance Spot

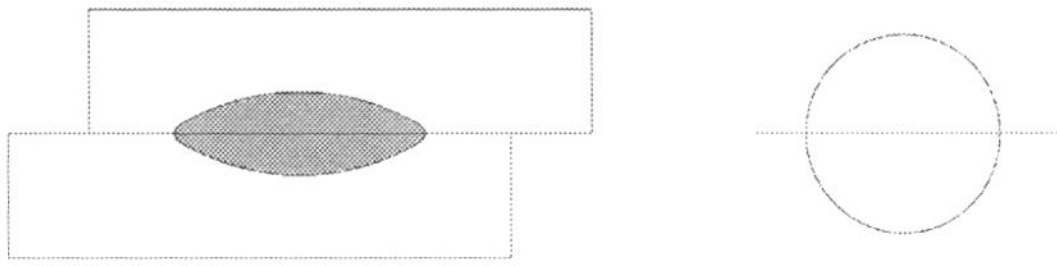

FIGURE 18.13 Resistance spot.

18.4.13 Projection

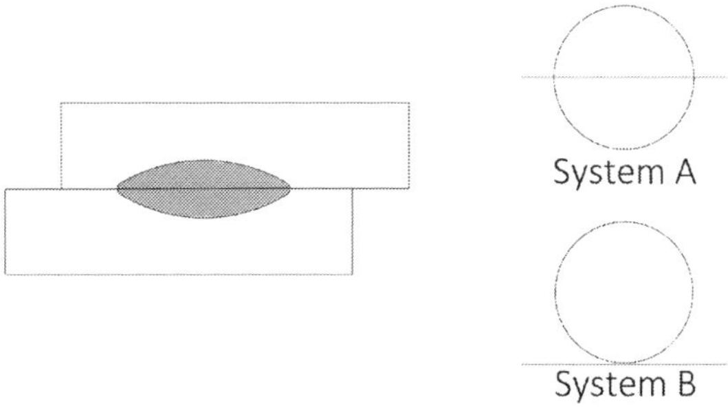

FIGURE 18.14 Projection.

18.4.14 Fusion Spot

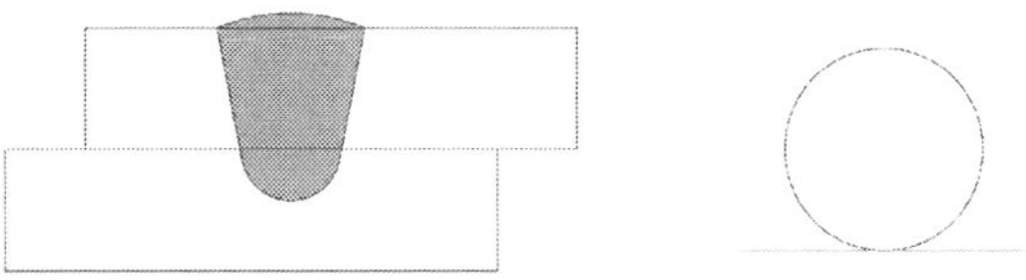

FIGURE 18.15 Fusion spot.

18.4.15 Resistance Seam

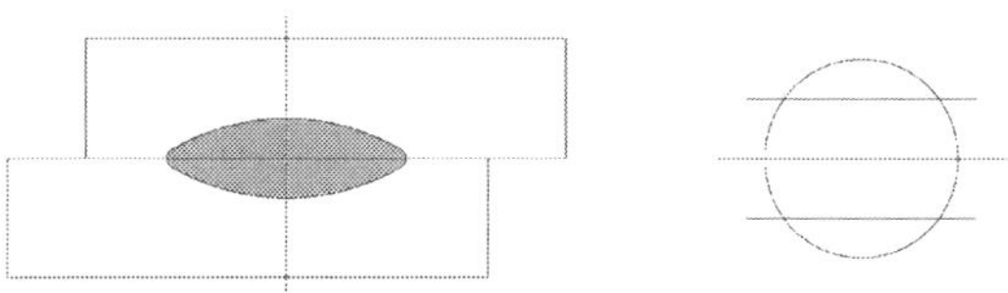

FIGURE 18.16 Resistance seam.

18.4.16 Fusion Seam

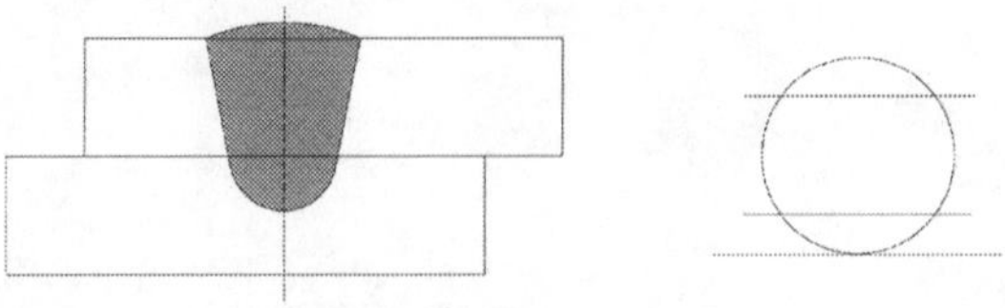

FIGURE 18.17 Fusion seam.

18.4.17 Stud

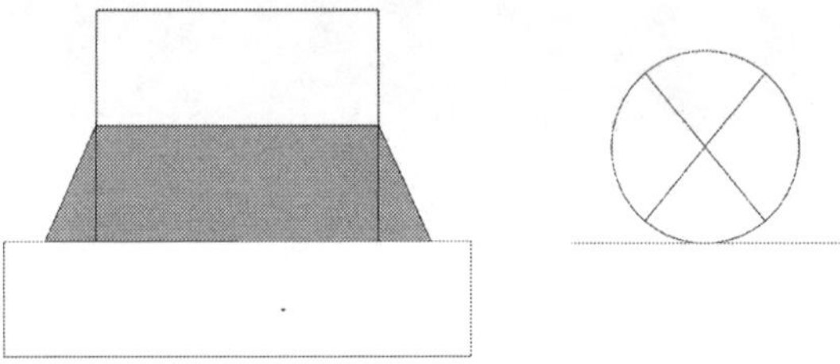

FIGURE 18.18 Stud.

18.4.18 Steep-Flanked Single-V Butt

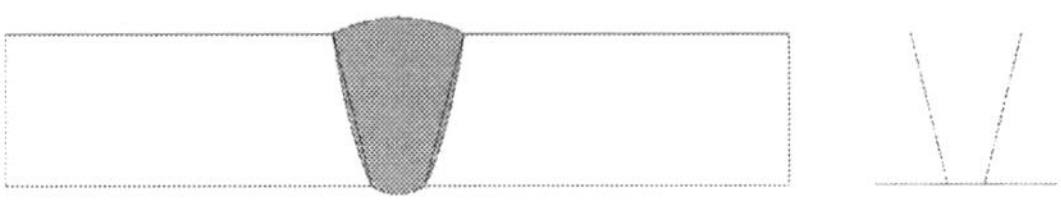

FIGURE 18.19 Steep-flanked single-V butt.

18.4.19 Steep-Flanked Single-Bevel Butt

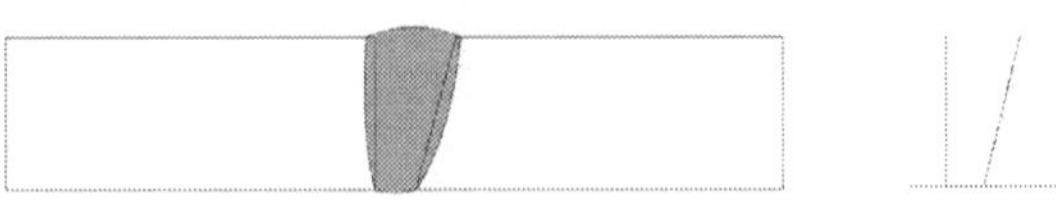

FIGURE 18.20 Steep-flanked single-bevel butt.

18.4.20 Edge

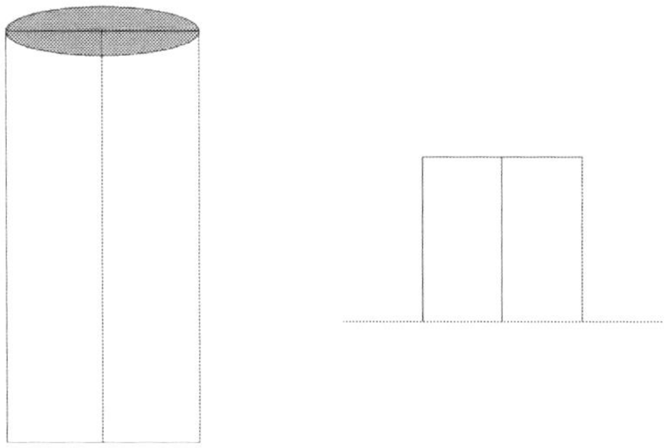

FIGURE 18.21 Edge.

18.4.21 Flanged Butt

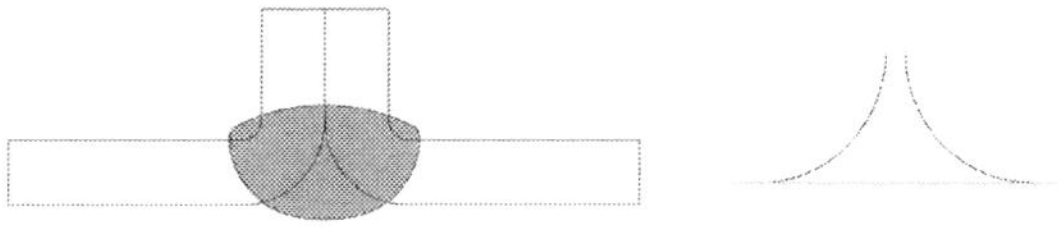

FIGURE 18.22 Flanged butt.

18.4.22 Flanged Corner

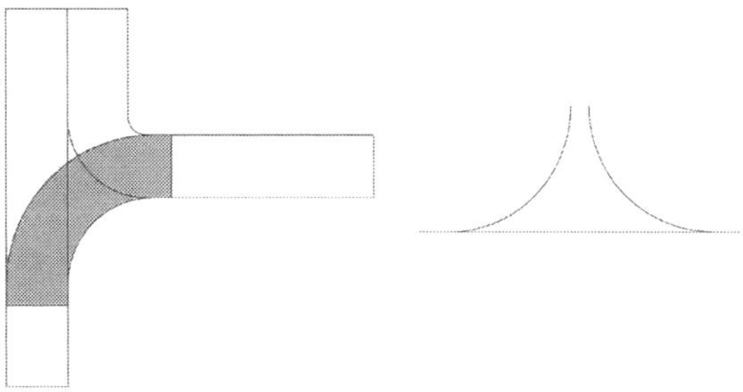

FIGURE 18.23 Flanged corner.

18.4.23 Overlay

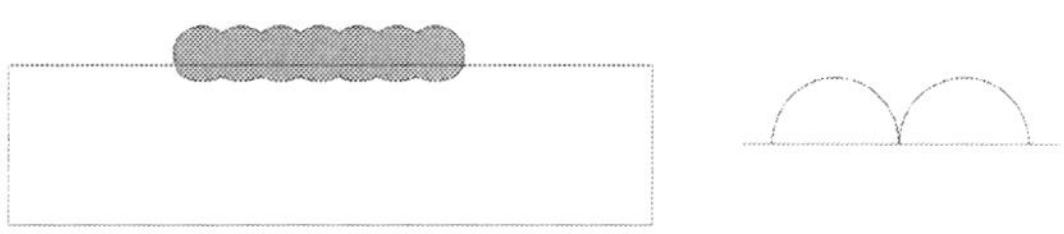

FIGURE 18.24 Overlay.

18.4.24 Stake

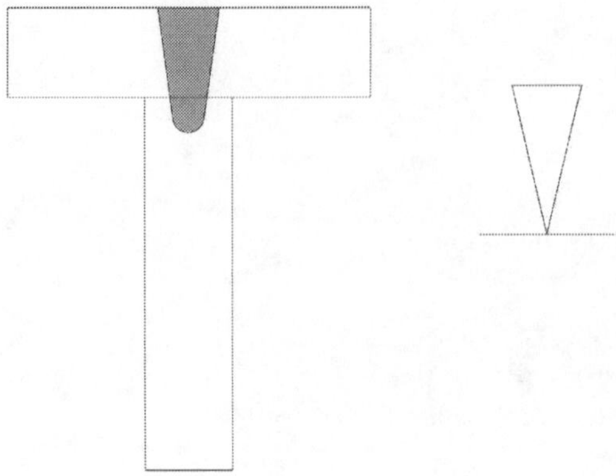

FIGURE 18.25 Stake.

18.5 SUPPLEMENTARY SYMBOLS

In addition to the elementary weld symbols, supplementary symbols can be added to clarify, for example, finishes on a weld. Common ones include those shown in Figure 18.26, but there are others defined in ISO 2553.

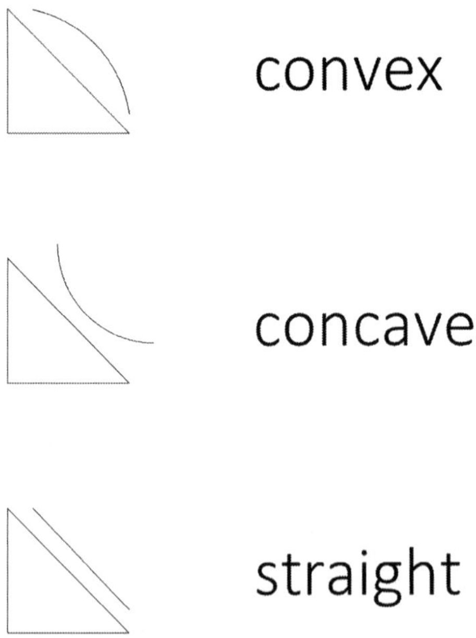

FIGURE 18.26 Example supplementary symbols.

18.6 EXAMPLE ELEMENTARY WELD

Figure 18.27 shows an example of a specification of a fillet weld using the line indicator. The letter 'z' indicates that the size of the weld should be measured on the vertical or 'leg length'. The weld should be concave in its outer shape, and there should be three welds, each 100 mm long and 100 mm apart (shown in brackets).

18.7 EXAMPLE COMBINED WELDS

Finally, Figure 18.28 shows how a combination of welds might be shown on an indicator for the common case where welding is carried out on both sides of a join, in this case, a double-V butt weld with broad root faces.

18.8 COMMENTS

Common weld symbols and their use in ISO standard drawings have been presented. The symbols are, for the most part, self-explanatory as to the shape of the cross-section of welds that would be produced. Note that some of the welds have gaps or cutaway areas that are to be filled in with filler materials while some rely entirely on melting/plasticizing/diffusing the basic substrate material.

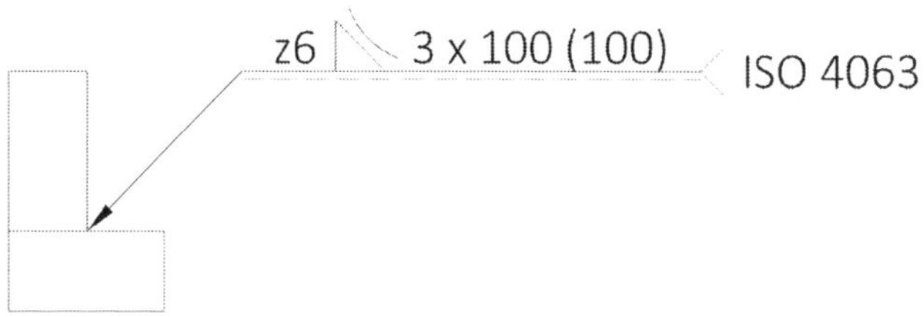

FIGURE 18.27 Example fillet weld indicator.

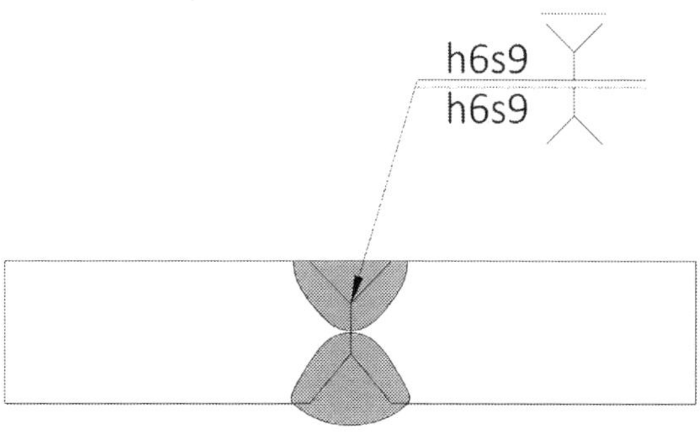

FIGURE 18.28 Double-V butt weld, preparation depth 6 mm weld 9 mm for both, top weld to be flat.

Index

0–9

1st angle, *see* first angle
3D solids models, 171
3rd angle, *see* third angle

A

absolute, 19, 21, 185, 187, 205
acceptance, 55, 201, 205
American National Standards Institution, *see* ANSI
American Society of Mechanical Engineers, *see* ASME
American Society for Testing and Materials, *see* ISO ASTM
American Welding Society, *see* AWS
angle plate, 91, 95–98, 112
ANSI (American National Standards Institution), xxi, 57, 67, 68, 69, 72
ANSI B1.1, 68
ANSI B1.5, 72
ANSI B1.15, 68
arc, 28–30, 33, 59, 128
ASME (American Society of Mechanical Engineers), xv, xxi–xxiii,14, 41, 43, 54, 67, 78, 80, 124
ASME 14.5 xxii–xxiii, 41, 43, 54, 80
AWS (American Welding Society) A2.4, 217

B

beta-spline, 138
Bezier, 31, 138–140, 143–144, 147–148, 152–153, 159, 161
BIM (Building Information Management), 171
bolt, 26, 53, 56–67, 71–72, 119
British Standards, *see* BS
broken view, 9–10
BS (British Standards), xvii–xxiv, 41, 57, 69–70
BS 21, 70
BS 164, 45
BS 817, 98
BS 8888, xxiii
b-spline, 31, 118, 138, 143–144, 146–147, 152–153, 159, 161, 171
Building Information Management, *see* BIM

C

calculated size, 34
Cartesian, 1, 3, 6, 136, 150, 169
chamfer, xxvii, 28–29, 71–72, 193
chord, 28–29, 148–149
chuck, 91, 97, 100, 180
circumscribed, minimum 37, 41, 91
CMM (Coordinate Measuring Machine), 80, 83, 85, 91, 94, 97, 127, 163, 180
combined zone, 86, 100–101, 118
cone, 1, 4, 45, 79, 85, 102, 129–131, 180
conical, 1–2, 4, 45, 127, 129–131, 147
constrained zone, 100
contract, 15–16, 43, 55, 162, 216
Coordinate Measuring Machine, *see* CMM
counterbore, 21, 26, 33
countersink, 26–27, 33
crank, 11–12, 66, 71
cross-section, 11, 35–36, 38, 65, 71–72, 76, 86–87, 108, 124, 129–131, 148, 154–156, 225
curve, xxv, xxvii, 29–31, 43, 86–88, 106, 115, 127, 130, 135–156, 160–161, 163, 171–172, 195, 203
cutaway, 11, 13, 21, 162, 174, 225
cutting plane, 10–11

D

datum, combined 100
datum, common 93, 100, 129–130, 133–134
datum, simulated 94–180
datum, system 82, 90–91, 93–94, 102, 126
datum, target 100–103
degrees of freedom, *see* DoF
Delaunay triangulation, 182–184
design intent, xxii, xxv, 15, 20, 34, 56, 67–68, 106, 135, 196
Deutsches Institut für Normung, *see* DIN
Digital Twin, xxi, 164, 166, 212, 216
dimension, 7, 17–19, 21, 24, 27–28, 31, 34, 36, 38, 41–43, 45, 49, 52, 84, 90, 152, 172
dimension, auxiliary 27–28, 80
dimension, chain 18–19
dimension, coordinate xxiii, 6, 20–22, 30–31, 80, 91, 136, 143, 146, 148–151, 160, 163, 169, 180, 185, 187, 192
dimension, half 23

dimension, imperial 6, 27–28, 58, 67, 70, 162, 173
dimension, metric xv, 6, 27–28, 56–59, 67–70, 160, 162
dimension, not-to-scale 27
dimension, ordinate 18–20, 115, 149, 151, 153
dimension, parallel xxii, 18–19
dimension, running 18
DIN (Deutsches Institut für Normung), 41, 69
DIN 876, 98
DIN 6063, 74
direct modelling, xxvi
discrete geometry, xxv, 140, 142, 205
DoF (degrees of freedom), 94, 96–97

E

edge, 28, 54, 58, 148–151, 155–156, 160, 165, 196–197, 215, 223
elevation, 2, 4, 11, 76, 102, 134
end elevation, 76
envelope, xxiii, 41–44, 83–85, 108–109, 117
Extensible Markup Language, *see* XML

F

fasteners, 177
feature
- associated 80, 91, 100, 102, 121, 125
- combined 86–87, 93–94, 100–101, 118
- datum 82, 85, 90–91, 96, 99–100, 103–104, 106, 108, 110, 112, 120, 125–126, 180
- derived 79, 91, 110, 120, 180–181
- extracted 80, 129, 180
- integral 34, 79, 85, 99, 110, 113, 180–181
- nominal 34, 80, 122, 126, 180
- real 36, 80, 91, 95
- situation 91–92, 109
- of size 34, 37–38, 40, 79, 81–82, 84, 120
- united 106, 117–118

fillet, xxvii,28–29, 160, 220, 225
first angle, 1–3, 6
fit
- clearance 46, 61
- close running 46, 53
- easy running 46
- interference 47, 61
- sliding 46
- transition 45–46, 52, 61

flat, 24–25, 32, 84, 91, 95–96, 98–100, 102, 106, 108, 110, 112, 120, 126, 128, 131–133, 180, 185, 190, 195, 225
flat plate, 98–99, 102
flexible, 9, 56, 100, 166, 169, 174

G

gauge dial, 42, 80, 83, 91, 93, 127, 129, 131–132, 178–179
gauge pins 42, 80, 91, 93
gauge, Go-NoGo 179
GD&T (Geometric Dimensioning and Tolerancing), xv, xxii, 16, 54, 78, 162
Geometric Dimensioning and Tolerancing, *see* GD&T
Geometric Product Specification, *see* GPS
geometric tolerance, xxii, 80, 100, 103–104, 115
GPS (Geometric Product Specification), xv, xxii–xxiii, xv, 16, 54, 78–80, 82, 85, 90, 103–104, 119, 151, 162, 178–179

H

hatching, 11, 13, 65, 134
history-based, xxvi
hole-based, 45, 47, 49, 52
hybrid modelling, xxvi

I

imperial, 6, 27–28, 57, 67, 70, 162, 173
independence, xxiii, 42–44, 54, 83, 86–87, 108–109, 129
indicator, additional 106–107, 215
indicator, adjacent 106, 108, 110, 112
inscribed, maximum 37
intermediate forms, 55
International Electrotechnical Commission, *see* ISO IEC
International Standards Organization, *see* ISO
Internet of Things, xxi, 164
ISO, GPS xv, xxii–xxii, 82, 85, 90, 103–104, 119, 151, 179
ISO (International Standards Organization), xv, xvii, xxi–xxiv, 14, 16, 30, 32–35, 40, 42–43, 45, 47, 52–54, 56–59, 61, 63, 67, 69–72, 74, 79–80, 82, 85–87, 90, 95, 106, 109, 115, 118–119, 151, 167, 179, 181, 189, 191–192, 194, 199, 204–205, 212–215, 217–218, 225
ISO 7–1, 70
ISO 14, 76
ISO 68, 58
ISO 228–1, 70
ISO 261, 58, 60
ISO 286, 34, 45, 47, 49
ISO 724, 59
ISO 841, 191
ISO 965, 59, 61–62, 64–66

ISO 1101, 54, 78, 80, 82, 89, 106, 116–118, 120, 173
ISO 1501, 69
ISO 1222, 57
ISO 2553, xxiv, 217, 224
ISO 2692, 90
ISO 2901, 71
ISO 2902, 71
ISO 2903, 71
ISO 2904, 71
ISO 4156, 76
ISO 4287, 196
ISO 4288, xxiv
ISO 5459, 90, 180
ISO 6413, 76
ISO 14405, 34–36, 38, 43, 78–80
ISO 16610, 184, 199
ISO 21920, xxiv, 196, 199–203, 205–208
ISO 25178, 196, 199, 206, 208
ISO ASTM 52900, 189
ISO IEC 30173, xxi, 216
ISO M series 56–58, 67
ISO R 1501, 69

J

Japanese Industrial Standards, *see* JIS
JIS (Japanese Industrial Standards), 69–70

K

keyway, 24–25
knurl, 55, 76, 77

L

laser scanning, xxiii, 80, 83
lathe, 45, 55–56, 70, 127
lead, 66, 71–72
legal, 15
limits and fits, 38, 45, 47, 53, 61, 76
line spacing, 13
line type, 13, 76, 212–214
local size, 34–38
long part, 9

M

macros, 21
MBD (Model-Based Definition), 166–167, 171
MBE (Model-Based Enterprise), 166–167
MBSE (Model-Based Systems Engineering), 166
metrology, xv, xxiii, 23, 37, 42, 78, 80, 83, 85, 97, 112, 151, 166, 178–179
mill, 24, 104, 185, 188
milled, 24
Model-Based Definition, *see* MBD
Model-Based Enterprise, *see* MBE
Model-Based Systems Engineering, *see* MBSE

N

non-rigid, 9, 32
NURBS, 31, 88, 138, 147–148, 153, 159, 161, 166–168, 171–172
nut, 56, 58–61, 64–69, 72

O

orthogonal, 1–4, 10, 15, 18

P

pairing, 49, 52–53
parameter, xxvi, 31, 38, 45, 53, 55, 58, 74, 138, 144–145, 148, 151–152, 160, 185, 196–210
parametric, xxvi, 136, 142, 144, 148, 152
partial views, 21–22
pattern, 7–8, 13, 21, 31, 56, 75–76, 123, 151, 193, 206, 214
PCD (pitch circle diameter), 7–8, 214
PDM (Product Data Management), 164–165
photo-realistic, 55
pitch circle diameter, *see* PCD
plan (view), 4, 10, 102
PLM (Product Lifecycle Management), 164–166
PMI (Product Manufacturing Information), xv, 16, 55–56, 74–75, 93, 99, 116, 165–167, 196, 199, 206–208, 212, 214–216
point-to-point, 19, 21, 178
prismatic, 11, 42, 91–92, 95
Product Data Management, *see* PDM
Product Lifecycle Management, *see* PLM
Product Manufacturing Information, *see* PMI
property indicator symbol, 32

Q

quadrant, 1–4, 147

R

rank order size, 34, 38
rendering, 55, 163, 178
right-hand rule, 170, 191
rounding error, xxv, 163–164, 182

S

SAE (Society of Automotive Engineers Standards), 69–70
screw, 26, 32, 45, 55–59, 67, 177
section view, 10–13, 174–175
shaft-based, 45, 47, 49, 53, 76
side elevation, 2, 4, 11, 102
Society of Automotive Engineers Standards, *see* SAE
special symbol, 31, 33
sphere, 24, 35–36, 79, 85, 91–92, 103, 116, 117, 119, 121, 153, 180
spherical, 24, 33, 35–36, 85, 102, 121
spline curve, 30–31, 135, 137–138, 143–153, 159, 161, 163, 171, 189, 195, 199
spline shaft, 33, 55, 57, 75–77
spring, 8–10, 13, 32, 174, 214
square section, 24
staggered, 11, 13
standard part, 26, 53, 55, 174, 177
statistical, 55, 179
subroutine, 21
surface, xxv, xxvii, 29–30, 56, 80, 82, 86, 88–89, 92, 106–107, 116-–121, 126–132, 135, 138, 147, 151–161, 163, 169, 172, 178, 180, 182, 189, 196
surface plate, 91, 93, 97–98
surface table xxiii, 95–99
surface texture xv, xxiv–xxv, 54, 196–210
surface treatment 59, 78
symmetry, 1, 5, 7, 11, 24, 120, 124–125, 213–214

T

taper, 23–24, 69–70
Technical Product Specification, *see* TPS
TED (Theoretically Exact Dimensions), 102–104, 114–115, 123, 126, 151
Theoretically Exact Dimensions, *see* TED
third angle, 1, 3–6, 11
thread
  acme 55, 70, 72–73
  angle 56, 58, 68
  BA 69
  BS 70
  chirality 66
  coarse 59, 60, 66, 68, 73
  crest 58–61, 64, 68, 71
  depth 58–59, 68
  engagement 65–66, 73
  fine 59–60, 66, 68–69, 73
  flank 58–62, 65
  handedness 66
  imperial 57
  left-handed 66, 72
  metric 69, 56
  parameters 53
  pipe 69–70, 74
  pitch 58–72, 76
  right-handed 66
  root 58–60, 68, 71
  square 55, 70–71
  trapezoidal 70–72
  UNC 57, 68
  Unified 67–68
  Whitworth 68–70
threaded, 33, 55–57, 61–63, 66, 70, 74–75, 119
through-hole, 21
tolerance
  angularity 44, 112–114
  concentricity 123
  flatness 84–85, 95, 112
  form 78, 81, 86, 88
  grade, 47–48, 61, 66, 69
  indicator 78, 83–87, 106–117, 119–120, 123–124, 151, 197, 202, 214–216
  line profile 86, 114–117, 120
  location(al) 103–104, 119, 120, 125, 128
  orientation 106–107, 114
  parallelism 106–110, 116, 120
  perpendicularity 110, 112
  position(al), 90, 105, 120, 124
  run-out, 82, 100, 125, 127–134
  straightness, 83–86
  surface profile, 88, 114–116, 120
  zone 23, 65, 84–88, 95, 100, 105–113, 115, 119, 121–134
TPS (Technical Product Specification), 78
turned, 45, 124–125, 134

V

validate, xxiii, 11, 17, 34–35, 42, 55, 83, 85, 89, 94–95, 98, 104, 106, 108, 110, 112, 128, 179, 180, 198, 201, 215
vee block, 42, 80, 83, 85, 97

X

XML (Extensible Markup Language), 166

Made in the USA
Las Vegas, NV
19 September 2026

b789b6b2-f40e-47fd-aa8c-2229c2515272R01